Rapid Assessment Program

A Biodiversity Assessment of Yongsu - Cyclops Mountains and the Southern Mamberamo Basin, Papua, Indonesia

Editors
Stephen J. Richards and Suer Suryadi

RAP
Bulletin
of Biological
Assessment
25

Center for Applied Biodiversity Science (CABS)

Conservation International

Indonesian Institute of Sciences (LIPI)

University of Cenderawasih (UNCEN)

The Papua Environment Foundation (YALI–Papua)

Dewan Masyarakat Adat Mamberamo Raya (DMAR)

The *RAP Bulletin of Biological Assessment* is published by:
Conservation International
Center for Applied Biodiversity Science
Department of Conservation Biology
1919 M St. NW, Suite 600
Washington, DC 20036
USA

202-912-1000 telephone
202-912-0773 fax
www.conservation.org
www.biodiversityscience.org

Editors: Stephen J. Richards and Suer Suryadi
Design/production: Kim Meek
Maps: Mark Denil
Translations: Suer Suryadi

RAP Bulletin of Biological Assessment Series Editors:
Terrestrial and Aqua RAP: Leeanne E. Alonso and Jennifer McCullough
Marine RAP: Sheila A. McKenna

ISBN: 1-881173-66-6
© 2002 by Conservation International
All rights reserved.
Library of Congress Catalog Card Number: 2002111859

Suggested citation:
Richards, S. J. and S. Suryadi (editors). 2002. A Biodiversity Assessment of Yongsu - Cyclops Mountains and the Southern Mamberamo Basin, Papua, Indonesia. RAP Bulletin of Biological Assessment 25. Conservation International, Washington, DC, USA.

Funding for the RAP training and surveys was provided by the MacArthur Foundation, USAID, Smart Family Foundation, Tropical Wilderness Protection Fund, and the Global Environmental Protection Institute. The Giuliani Family Foundation supported publication of this report.

Table of Contents

Participants 7

Organizational Profiles 10

Acknowledgments 12

Report At A Glance 13

Laporan Ringkas 15

Executive Summary 21

Ringkasan Eksekutif 26

Chapter 1 32
Geographic overview of the Cyclops Mountains and the
Mamberamo Basin

Dan A. Polhemus and Stephen Richards

Bab I 38
Ulasan Geografis Pegunungan Cyclops dan Daerah
Aliran Sungai Mamberamo

Dan A. Polhemus dan Stephen Richards

Chapter 2 45
Plant diversity in lowland forests of the Yongsu area,
Papua, Indonesia

*Yance de Fretes, Conny Kameubun, Ismail A. Rachman,
Julius D. Nugroho, Elisa Wally, Herman Remetwa,
Marthen Kabiay, Ketut G. Suartana, and Basa T.
Rumahorbo*

Chapter 3 51
Vegetation of the Dabra area, Mamberamo River Basin,
Papua, Indonesia

Yance de Fretes, Ismail A. Rachman, and Elisa Wally

Chapter 4 57
Aquatic Insects of the Dabra area, Mamberamo River
Basin, Papua, Indonesia

Dan A. Polhemus

Chapter 5 61
Butterflies of the Yongsu area, Papua, Indonesia

*Edy Rosariyanto, Henk van Mastrigt, Henry Silka Innah,
and Hugo Yoteni*

Chapter 6 63
Butterflies and Moths of the Dabra area, Mamberamo
River Basin, Papua, Indonesia

Henk van Mastrigt and Edy M. Rosariyanto

Chapter 7 67
Fishes of the Yongsu and Dabra areas, Papua, Indonesia

*Gerald R. Allen, Henni Ohee, Paulus Boli, Roni Bawole,
and Maklon Warpur*

Chapter 8 73
Amphibians and Reptiles of the Yongsu area, Papua,
Indonesia

*Stephen Richards, Djoko T. Iskandar, Burhan Tjaturadi,
and Aditya Krishar*

Chapter 9 76
Amphibians and Reptiles of the Dabra area, Mamberamo
River Basin, Papua, Indonesia

*Stephen Richards, Djoko T. Iskandar, and Burhan
Tjaturadi*

Chapter 10 80
Birds of the Yongsu area, northern Cyclops Mountains, Papua, Indonesia

Pujo Setio, Paul Johan Kawatu, David Kalo, Daud Womsiwor, and Bruce M. Beehler

Chapter 11 84
Birds of the Dabra area, Mamberamo River Basin, Papua, Indonesia

Bas van Balen, Suer Suryadi, and David Kalo

Chapter 12 89
Small mammals of the Dabra area, Mamberamo River Basin, Papua, Indonesia

Rose Singadan and Freddy Patiselanno

Gazetteer of Focal Sites, Yongsu and Mamberamo 92

Appendices 93

Appendix 1 96
Seedlings and grasses (≤ 50 cm in height) sampled in 10 sub-plots of 0.5 x 2 m in each of five 0.1 ha plots in Jari forest, Yongsu Dosoyo, Papua, Indonesia

Yance de Fretes, Conny Kameubun, Ismail A. Rachman, Julius D. Nugroho, Elisa Wally, Herman Remetwa, Marthen Kabiay, Ketut G. Suartana, and Basa T. Rumahorbo

Appendix 2 98
Saplings (1.0–4.9 cm dbh) in 2 sub-plots of 2 x 5 m in each of five 0.1 ha plots in Jari forest, Yongsu Dosoyo, Papua, Indonesia

Yance de Fretes, Conny Kameubun, Ismail A. Rachman, Julius D. Nugroho, Elisa Wally, Herman Remetwa, Marthen Kabiay, Ketut G. Suartana, and Basa T. Rumahorbo

Appendix 3 100
Poles (trees 5.0–9.9 cm dbh) sampled in 5 sub-plots of 5 x 20 m in each of five 0.1 ha plots in Jari forest, Yongsu Dosoyo, Papua, Indonesia

Yance de Fretes, Conny Kameubun, Ismail A. Rachman, Julius D. Nugroho, Elisa Wally, Herman Remetwa, Marthen Kabiay, Ketut G. Suartana, and Basa T. Rumahorbo

Appendix 4 101
Trees (dbh ≥ 10 cm) sampled in 5 sub-plots of 20 x 50 m and 5 sub-plots of 5 x 20 m in each of five 0.1 ha plots in Jari forest, Yongsu Dosoyo, Papua, Indonesia

Yance de Fretes, Conny Kameubun, Ismail A. Rachman, Julius D. Nugroho, Elisa Wally, Herman Remetwa, Marthen Kabiay, Ketut G. Suartana, and Basa T. Rumahorbo

Appendix 5 104
Beach vegetation observed around Yemang (Yongsu) Training Camp, Papua, Indonesia

Yance de Fretes, Conny Kameubun, Ismail A. Rachman, Julius D. Nugroho, Elisa Wally, Herman Remetwa, Marthen Kabiay, Ketut G. Suartana, and Basa T. Rumahorbo

Appendix 6 105
Vegetation recorded from secondary forests around Yemang Training Camp, Yongsu Dosoyo, Papua, Indonesia

Yance de Fretes, Conny Kameubun, Ismail A. Rachman, Julius D. Nugroho, Elisa Wally, Herman Remetwa, Marthen Kabiay, Ketut G. Suartana, and Basa T. Rumahorbo

Appendix 7 106
Vegetation recorded outside of plots during surveys of forests at Jari and Yemang, Yongsu area, Papua, Indonesia

Yance de Fretes, Conny Kameubun, Ismail A. Rachman, Julius D. Nugroho, Elisa Wally, Herman Remetwa, Marthen Kabiay, Ketut G. Suartana, and Basa T. Rumahorbo

Appendix 8 109
Grasses, seedlings, and herbaceous plants (≤ 50 cm in height) in 50 subplots (0.5 x 2 m) in five 20 x 50 m Whittaker Plots at Furu River, Papua, Indonesia

Yance de Fretes, Ismail A. Rachman, and Elisa Wally

Appendix 9 111
List of all trees (≥ 1 cm dbh) sampled in 5 Whitaker plots (Furu and Jari) and 2 transects (Tiri) during 2000 RAP surveys, Papua, Indonesia

Yance de Fretes, Ismail A. Rachman, and Elisa Wally

Appendix 10 119
Botanical specimens collected outside of plots during Papua (Indonesia) RAP surveys (2000) and identified by Ismail A. Rachman

Yance de Fretes, Ismail A. Rachman, and Elisa Wally

Appendix 11 127
Plant species grown in gardens and used by the Dabra community (Mamberamo Basin), Papua, Indonesia

Yance de Fretes, Ismail A. Rachman, and Elisa Wally

Appendix 12 129
Sampling stations for aquatic insect surveys in the Dabra area, Papua, Indonesia

Dan A. Polhemus

Appendix 13 130
Aquatic insects captured at nine sampling stations in the Dabra area, Papua, Indonesia

Dan A. Polhemus

Appendix 14 133
Annotated checklist of aquatic insects collected during the Dabra RAP survey, Papua, Indonesia.

Dan A. Polhemus

Appendix 15 137
Aquatic insects collected in the Cyclops Mountains, Papua, Indonesia

Dan A. Polhemus

Appendix 16 138
List of butterflies collected around Yongsu Dosoyo, Papua, Indonesia

Edy Rosariyanto, Henk van Mastrigt, Henry Silka Innah, and Hugo Yoteni

Appendix 17 140
List of butterflies recorded at Furu and Tiri Rivers, Mamberamo Basin, Papua, Indonesia

Henk van Mastrigt and Edy M. Rosariyanto

Appendix 18 142
Diversity of moths collected in the Dabra area, Papua, Indonesia

Henk van Mastrigt and Edy M. Rosariyanto

Appendix 19 144
Summary of fish collection/observation sites in the Yongsu and Dabra areas, Papua, Indonesia

Gerald R. Allen

Appendix 20 146
Summary of freshwater fishes collected during the Yongsu training course, Papua, Indonesia

Gerald R. Allen

Appendix 21 148
Annotated checklist of fishes of the Yongsu area, Papua, Indonesia

Gerald R. Allen

Appendix 22 151
List of shallow coral reef fishes of Yongsu Bay, Papua, Indonesia

Gerald R. Allen

Appendix 23 155
Summary of fishes collected on the RAP survey in the Mamberamo River drainage, Papua, Indonesia

Gerald R. Allen

Appendix 24 157
Annotated checklist of fishes recorded from the Mamberamo River system, Papua, Indonesia

Gerald R. Allen

Appendix 25 160
List of fish species recorded to date from the Mamberamo, Sepik, and Ramu Rivers of northern New Guinea

Gerald R. Allen

Appendix 26 162
Distribution of amphibians and reptiles in the Yongsu area (0–400 m asl), Papua, Indonesia

Stephen Richards, Djoko Iskandar, Burhan Tjaturadi, and Aditya Krishar

Appendix 27 164
Frogs and reptiles recorded from three sites in the Dabra area, Papua, Indonesia

Stephen Richards, Djoko Iskandar, and Burhan Tjaturadi

Appendix 28 166
Annotated list of noteworthy frogs and reptiles recorded from three sites in the Dabra area, Papua, Indonesia

Stephen Richards, Djoko Iskandar, and Burhan Tjaturadi

Appendix 29 168
Birds recorded at Yongsu, Papua, Indonesia

Pujo Setio, Paul Johan Kawatu, Irba U. Nugroho, David Kalo, Daud Womsiwor, and Bruce M. Beehler

Appendix 30 71
Forest trees at two bird census points, Yongsu, Papua, Indonesia

Pujo Setio, Paul Johan Kawatu, Irba U. Nugroho, David Kalo, Daud Womsiwor, and Bruce M. Beehler

Appendix 31 172
Birds recorded from the Mamberamo/Idenburg river
basins, Papua, Indonesia

Bas van Balen, Suer Suryadi, and David Kalo

Appendix 32 179
Annotated list of noteworthy birds known from or expected
to occur in the Dabra area

Bas van Balen, Suer Suryadi, and David Kalo

Participants

Gerald R. Allen (Yongsu and Mamberamo, Fish)
Conservation International
1 Dreyer Road
Roleystone, WA 6111
AUSTRALIA
Email: tropical_reef@bigpond.com

Roni Bawole (Yongsu, Fish)
Universitas Negeri Papua
Jl. Gunung Salju Amban
Manokwari 98314, Papua
INDONESIA
Email: incune@manokwari.wasantara.net.id

Bruce Beehler (Yongsu, Birds)
Conservation International
1919 M Street, NW, Suite 600
Washington, DC 20036
USA
Email: b.beehler@conservation.org

Paulus Boli (Yongsu and Mamberamo, Fish)
Universitas Negeri Papua
Department of Fishery and Marine Science
Jl. Gunung Salju Amban
Manokwari 98314, Papua
INDONESIA
Email: incune@manokwari.wasantara.net.id

Yance de Fretes (Yongsu and Mamberamo, Coordination
    and Vegetation)
Conservation International-Indonesia
Jl. Taman Margasatwa No. 61
Jakarta 12540
INDONESIA
Email: Yance@conservation.or.id

Debbie Gowensmith (Yongsu and Mamberamo,
    Coordination and Reporting)
Conservation International
Rapid Assessment Program
1919 M Street NW, Suite 600
Washington, DC 20036
USA

Henry Silka Innah (Yongsu, Butterflies)
BPK Manokwari
Jl. Inamberi Pasir Putih
P.O. Box 159
Manokwari, Papua 98301
INDONESIA
Email: bpk_mkw@manokwari.wasantara.net.id

Djoko T. Iskandar (Yongsu and Mamberamo,
    Herpetofauna)
Institut Teknologi Bandung
Jurusan Biologi
Lab. Tek. XI ITB
Jl. Ganesa 10
Bandung 40132
INDONESIA
Email: iskandar@bi.itb.ac.id

Marthen Helios Kabiay (Yongsu, Vegetation)
YALI-Jayapura
Jl. Tobati No. 2
Jayapura, Papua
INDONESIA

David Kalo (Yongsu and Mamberamo, Birds)
Balai Konservasi Sumber Daya Alam Irian Jaya I Jayapura
Jl. Raya, Abepura, Kotaraja
Jayapura, Papua
INDONESIA

Conny Kameubun
Conservation International-Papua
Jl. Bhayangkara I No. 33
Jayapura 99112, Papua
INDONESIA

Paul Johan Kawatu (Yongsu, Birds)
UNCEN-Jayapura
Kampus F. MIPA Biologi UNCEN
Jl. Kampus Baru UNCEN Waena
Perumnas III Jayapura, Papua
INDONESIA

Aditya Krishar (Yongsu, Herpetofauna)
UNCEN-Jayapura
Kampus F. MIPA Biologi UNCEN
Jl. Kampus Baru UNCEN Waena
Perumnas III Jayapura, Papua
INDONESIA

Irba Unggul Nugroho (Yongsu, Birds)
Universitas Negeri Papua
Jl. Gunung Salju Amban
Manokwari 98314, Papua
INDONESIA
Email: incune@manokwari.wasantara.net.id

Julius Nugroho (Yongsu, Vegetation)
Universitas Negeri Papua
Jl. Gunung Salju Amban
Manokwari 98314, Papua
INDONESIA
Email: incune@manokwari.wasantara.net.id

Henni Ohee
Conservation International-Papua
Jl. Bhayangkara I No. 33
Jayapura 99112, Papua
INDONESIA

Zeth Parinding (Yongsu, Mammals)
Taman National Wasur
Jl. Raya Mandala Spadem No. 2
Merauke, Papua
INDONESIA

Freddy Pattiselanno (Yongsu and Mamberamo, Mammals)
Universitas Negeri Papua
Jl. Gunung Salju Amban
Manokwari 98314, Papua
INDONESIA
Email: incune@manokwari.wasantara.net.id

Dan A. Polhemus (Mamberamo, Aquatic Insects)
Department of Entomology
MRC 105
Smithsonian Institution
Washington, DC 20560
USA
Email: bugman@mail.bishopmuseum.org

Ismail A. Rachman (Yongsu and Mamberamo, Vegetation)
Balitbang Botani
Puslitbang Biologi-LIPI
Jl. Ir. H. Juanda 22
Bogor 16122
INDONESIA
Email: herbogor@indo.net.id

Herman Remetwa (Yongsu, Vegetation)
BPK Manokwari
Jl. Inamberi Pasir Putih
P.O. Box 159
Manokwari, Papua 98301
INDONESIA

Stephen Richards (Yongsu and Mamberamo, Team Leader
   and Herpetofauna)
Vertebrate Department
South Australian Museum
North Terrace
Adelaide, S.A. 5000
AUSTRALIA
Email: SJRichardsEnvtl@bigpond.com

Edy Michelis Rosariyanto (Yongsu and Mamberamo,
   Butterflies)
Conservation International-Papua
Jl. Bhayangkara I No. 33
Jayapura 99112, Papua
INDONESIA
Email: ci-irian@jayapura.wasantara.net.id

Basa T. Rumahorbo (Yongsu, Vegetation)
UNCEN-Jayapura
Kampus F. MIPA Biologi UNCEN
Jl. Kampus Baru UNCEN Waena
Perumnas III Jayapura, Papua
INDONESIA

Pujo Setio (Yongsu, Birds)
BPK Manokwari
Jl. Inamberi Pasir Putih
P.O. Box 159
Manokwari, Papua 98301
INDONESIA
Email: bpk_mkw@manokwari.wasantara.net.id

Rose Singadan (Yongsu and Mamberamo, Mammals)
Biology Department
School of Natural and Physical Sciences
University of Papua New Guinea
P.O. Box 320 University
PAPUA NEW GUINEA
Email: rose.singadan@upng.ac.pg

Ketut G. Suartana (Yongsu, Vegetation)
BKSDA II-Sorong
Jl. Jend. Sudirman No. 40
Sorong, Papua
INDONESIA

Suer Suryadi (Yongsu and Mamberamo, Coordination
    and Birds)
Conservation International-Papua
Jl. Bhayangkara I No. 33
Jayapura 99112, Papua
INDONESIA
Email: suer@conservation.or.id

Burhan Tjaturadi (Yongsu and Mamberamo, Herpetofauna)
WWF Bio Region Sahul
Jl. Angkasa Indah II/No. 10
Jayapura, Papua
INDONESIA
Email: Btjaturadi@wwfnet.org

Bas van Balen (Mamberamo, Birds)
Tropical Nature Conservation & Vertebrate Ecology Group
Wageningen University
Bornsestug 69 Wageningen
THE NETHERLANDS
Email: bas.vanbalen@staf.ton.wau.nl

Henk van Mastrigt (Yongsu and Mamberamo, Butterflies)
Kotak Pos 1078
Jayapura 99010
INDONESIA
Email: henkvm@jayapura.wasantara.net.id

Elisa Wally (Yongsu and Mamberamo, Vegetation)
Pusat Studi Keanekaragaman Hayati
(The Biodiversity Study Center)
Universitas Negeri Papua
Manokwari 98314, Papua
INDONESIA
Email: incune@manokwari.wasantara.net.id

Maklon Warpur (Yongsu, Fish)
UNCEN-Jayapura
Kampus F. MIPA Biologi UNCEN
Jl. Kampus Baru UNCEN Waena
Perumnas III Jayapura, Papua
INDONESIA

Daud Womsiwor (Yongsu, Birds)
BKSDA II-Sorong
Jl. Jend. Sudirman No. 40
Sorong, Papua
INDONESIA

Hugo Yoteni
Conservation International-Papua
Jl. Bhayangkara I No. 33
Jayapura 99112, Papua
INDONESIA

Organizational Profiles

Conservation International

Conservation International (CI) is an international, non-profit organization based in Washington, DC. CI acts on the belief that the Earth's natural heritage must be maintained if future generations are to thrive spiritually, culturally, and economically. Our mission is to conserve biological diversity and the ecological processes that support life on earth and to demonstrate that human societies are able to live harmoniously with nature.

Conservation International
1919 M Street NW, Suite 600
Washington DC, 20036, USA
Tel: 202-912-1000
Fax: 202-912-1046
http://www.conservation.org

Conservation International-Papua

CI-Papua was formally established in 1995, after a CI analysis designated the island of New Guinea as a key Major Tropical Wilderness Area. CI's long-term goal in Papua is to promote people-based nature conservation and sustainable economic development through the sustainable use of Papua's resources. The strategy for the province is based on the recognition that Papua still has large, virtually unexplored areas of high biodiversity that are globally significant and should be targeted for in-depth scientific study. To help meet the conservation and development needs of the province, CI is pursuing a long-term strategy to assist Indonesian institutions and NGOs in tackling the challenge of designing and implementing conservation and sustainable development initiatives in Papua.

CI Papua   Mailing address:
Jl. Bhayangkara I No. 33   P.O. Box 344
Jayapura 99112   Jl. Sentani, No. 1
Papua   Abepura, Jayapura, 99351, Papua
INDONESIA   INDONESIA
Tel/Fax: 0967-523423
E-mail: ci-irian@jayapura.wasantara.net.id
http://www.conservation.or.id

Conservation International-Indonesia

CI-Indonesia was founded in 1992 to protect and conserve biodiversity in the hotspot regions of Indonesia that include the Eastern Sundaic Region covering East Nusa Tenggara, West Nusa, Tenggara, Sulawesi, and Maluku; and the Western Sundaic Region that includes Sumatra, Java, Kalimantan, Bali, and Lombok Island. The Tropical Wilderness Area of Papua is also a focus of the program. CI-Indonesia's approach is to offer technical support to Indonesian institutions and to facilitate the establishment of long-term field projects that promote community-based conservation and sustainable development. CI also emphasizes the scientific assessment of biodiversity and the documentation of economic, cultural, and social importance of biodiversity for Indonesian people. Through collaborations with many Indonesian institutions, CI facilitates discussion among stakeholders at national, regional, and local levels to make strategic decisions regarding where protected areas should be located and to come to an understanding of the importance of conserving biodiversity in Indonesia.

Conservation International-Indonesia
Jl. Taman Margasatwa No. 61
Jakarta 12540, INDONESIA
Tel: 62-21-7883-8624 or 62-21-7883-8626
Fax: 62-21-780-0265
Email: CI-Indonesia@conservation.org
http://www.conservation.or.id

Indonesian Institute of Sciences (LIPI)

The Indonesian Institute of Sciences (LIPI) is a non-departmental institution that reports directly to the President of Indonesia. The main tasks of LIPI are to assist the President in organizing research and development, and to provide guidance, services, and advice to the government on national science and technology policy. In order to accomplish its main tasks, LIPI was assigned the following functions:

1. To carry out research and development of science and technology.
2. To encourage and develop science consciousness among the Indonesian people.
3. To develop and improve cooperation with national as well as international scientific bodies in accordance with the existing laws and regulations.
4. To provide the government with the formulation of national science policy.

Puslitbang Biologi–LIPI
(Research and Development Center for Biology)
Jl. Juanda 18
Bogor, Jawa Barat 16004
INDONESIA

University of Cenderawasih (UNCEN)

University of Cenderawasih is based in Jayapura, Papua, and serves as the center of educational and research excellence for Papuan students. UNCEN's mission has been to train and enhance the technological and human resources of Papua to the benefit of the Papuan people and community. Focus has been on law, economics, education, natural sciences, and social politics.

University of Cenderawasih
UNCEN-Jayapura
Jl. Kampus Baru UNCEN Waena
Perumnas III
Jayapura, Papua
INDONESIA

The Papua Environment Foundation (YALI–Papua)

YALI (Yayasan Lingkungan Hidup)–Papua is a non-profit organization founded in Jayapura in 1994. YALI's vision is to protect and conserve ecological systems and to assist people to access and use natural resources in a sustainable way. To implement the program, YALI always considers the balance between exploitation of natural resources, recognition of indigenous rights, professionalism, and sustainable resource management.

YALI-Papua
Jl. Ifar No. 62, Kel. Asano,
Abepura 99351
Jayapura, Papua
INDONESIA
Email: bakau@jayapura.wasantara.net.id

Dewan Masyarakat Adat Mamberamo Raya (DMAR)

DMAR (Great Mamberamo Adat Council) is a community-based organization established in 1998. The purpose of the Council is to protect rights to their lands, and to use their forests and rivers to improve their lives. In doing this they hope to consolidate and empower their community, and to represent and facilitate community expectations in terms of sustainable and integrated development, including economic and social-culture development.

DMAR
Jl Tobati No 81, Kel Asano,
Abepura 99351
Papua
INDONESIA

Acknowledgments

The success of this RAP program was due to the tremendous support of many people. The participants thank Suer Suryadi of CI-Papua for his hard work and exceptional organising skills, which he used to solve many problems and which ultimately ensured the success of the training and expedition. Leeanne Alonso was instrumental in bringing the RAP to fruition and was the backbone of the project in Washington. This was a joint CI and University of Cenderawasih (UNCEN) program and we are most grateful to Mr. Frans Wospakrik and Mr. Sam Renyaan of UNCEN for their considerable input, and for their assistance with obtaining the provincial permits required. Dr. Siti N. Prijono, Dr. Arie Budiman, and Dr. Deddy Darnaedi of LIPI kindly helped us to obtain the permits required from Jakarta for this biological assessment. We would also like to extend our gratitude to the head of BKSDA Jayapura, I.G. Sutedja, for his tremendous support of our CI-Indonesia programs and to Pak Wouter Sahanaya of USAID–Jakarta for his support of CI's activities in Papua.

Robert Mandosir (Director of YALI), Abner Mansay (Vice Director), and Nico Wamafma (YALI-Papua) facilitated all of our interactions and collaborations with local communities and in particular with local stakeholders in the project. We greatly appreciated their knowledge and friendship before and after the expeditions. Wempi Bilasy of Mamberamo Adat Council and his staff introduced us to the people of Mamberamo and Dabra, supported us in the field, and encouraged us to do our best for the Mamberamo region. Dr. Jatna Supriatna, Sari Surjadi, Ermayanti, Iwan Wijayanto, Muhammad Farid, Hendritte Ohee, Mira Dwiarsanty, Myrna Kusumawardhani, Yuli Nurhayati, Hugo Yoteni, Yehuda Demetou, and Max Ormuseray of CI-Indonesia worked incredibly hard for a long period of time and often under difficult circumstances to organise the complex logistics required for a project like this. Debbie Gowensmith assisted with numerous facets of the project, both in the field and in the office, and her unfailing good humour under uncomfortable field conditions was very much appreciated.

Other organizations that supported this program in different ways include: BKSDA-Jayapura, Yajasi, Tariku, and AMA airlines. The Church organizations GIDI and GKI provided letters of recommendation for conducting this training and expedition.

We would like to express our sincere thanks to the community leaders (Ondoafi of Dabra and Ondoafi of Yongsu Dosoyo) and to our excellent local guides: Yongsu Dosoyo - Zet Ormuseray, Daniel Yoafifi, Kalvin Nari, Yordan Ormuseray, Benyamin Abisay, Eli Ormuseray, Eliezer Okoseray, Simson Nusa, Melkisedek Tablaseray, and Simon Ormuseray; Dabra-Mamberamo - Musa Abaiso, Benyamin Foisa, Waenan Fruaro, Markus Sarife, Klaus Foisa, Barnabas Fruaro, Onesmus Abaiso, Agus Foisa, Konstan Abaiso, Leuwi, Obet Foisa, Evert Biso, Yohannes Biso, and Otis Foisa. Tony Seroyer of Mamberamo Hulu sub-district office and the Merpati ground-crew in Dabra were very supportive and helpful. We thank all of these people and the many other community members from Yongsu and Dabra who helped to make this training and expedition successful.

We thank the following sponsors for providing the financial support for the RAP training and expedition: the MacArthur Foundation, USAID, Smart Family Foundation, Tropical Wilderness Protection Fund, and the Global Environmental Protection Institute. This publication was made possible by the generous support of The Giuliani Family Foundation.

Report At A Glance

A BIODIVERSITY ASSESSMENT OF YONGSU - CYCLOPS MOUNTAINS AND THE SOUTHERN MAMBERAMO BASIN, PAPUA, INDONESIA

Dates of Expedition
Yongsu RAP Training Course: August 19–30, 2000
Mamberamo RAP Expedition: September 1–15, 2000

Description of Expedition
The RAP training course took place in the vicinity of Yongsu Dosoyo on the northern edge of the Cyclops Mountains Nature Reserve, Papua province, Indonesia. The region has high topographic relief with mountain ridges plunging from more than 1000 m elevation to sea level in a distance of less than 5 km. There is no coastal plain and training was conducted in a narrow strip of forest sandwiched between the steep northern slopes of the Cyclops Mountains and the ocean.

The RAP Expedition explored a wide variety of aquatic and terrestrial habitats at two sites near the village of Dabra in the southern Mamberamo River Basin. The basin supports a vast area of pristine lowland rainforest on the northern side of Papua's central cordillera. It is the largest catchment in Papua, draining all northward flowing streams that descend from the central mountains between the Papua New Guinea border and approximately 137° west longitude. Water levels in the main river fluctuate dramatically through the year, creating a variety of aquatic habitats including swampy and flooded forest, swampy grasslands, oxbows, and small lakes. There is a transition from lowland to foothill forest at the southern edge of the basin where the central cordillera rises steeply from the swampy lowlands. Our survey was conducted during the dry season when water levels in the main river channel and in the streams draining the central cordillera were relatively low.

Reason for RAP Survey and Training
The training course was conducted because Papua has a critical shortage of scientists with the skills to rapidly collect, analyse and disseminate biodiversity information that is critical for making well-informed conservation recommendations. The course also provided the opportunity to survey the flora and fauna of the northern coastal fringe of the Cyclops Mountains.

The Mamberamo Basin was the focus for the RAP expedition because it is a vast area of pristine, sparsely populated forest that has been poorly documented and is faced with increasing threats. The proposed Mamberamo Mega-Project includes an aluminium smelter and extensive agro-industry in the basin, powered by a hydro-electric dam that will flood a substantial area of forest. The dam would also modify water flow patterns downstream with serious impacts on ecological dynamics of the forest and aquatic ecosystems. The influx of people associated with the project would undoubtedly have serious cultural impacts on the indigenous inhabitants of the basin. If this project proceeds it will have catastrophic impacts on the basin which is already facing increased pressure from logging operations. Under these conditions assessments of the current health and biodiversity values of ecosystems in this vast wilderness area are urgently required to assist with the development of appropriate conservation and development strategies.

Major Results

Twenty three local scientists from the University of Cenderwasih, NGOs, and agencies from the forestry department were trained by 10 national and international scientists each with expertise in the flora or fauna of the region. Forests of the Yongsu area provide important resources for local communities but current rates of plant and animal harvesting appear to be sustainable and the forests remain largely intact away from the immediate vicinity of villages. This contrasts with other areas on the northern fringe of the Cyclops Mountains where forests have been severely degraded. Yongsu communities are supporting conservation activities in the area so the long-term survival of healthy forests in the region seems to be viable.

The Mamberamo survey found evidence that the forests, aquatic ecosystems, and river soils in the Dabra region are in good health. Human population levels are still very low and the total area converted to agriculture is small. However, from the air the encroachment of logging roads into the basin from the north was clearly evident. Exotic fish represented an alarming proportion (17.1%) of the fish documented during the survey and their impact on the aquatic ecosystems and on native fish species should be assessed as a matter of urgency. Despite their proximity, the two main sites surveyed (Furu and Tiri) had slightly different floras and faunas. There was more secondary vegetation at Furu than at Tiri as a result of greater human impact at that site. Furthermore the presence of hill forest only at the Furu site provided access to a different suite of plants and animals there. Additional surveys within the basin at a range of altitudes are required before a comprehensive assessment of its biodiversity values can be attempted.

Number of Species (ns = not surveyed)

	Yongsu	Mamberamo
Plants:	178	*Tiri:* 131
		Furu: 234
Aquatic insects:	ns	56
Butterflies:	69	129
Freshwater fish:	33	23
Marine reef fish:	195	ns
Frogs:	8	21
Reptiles:	26	36
Birds:	90	143
Mammals:	ns	7

New Species Discovered

	Yongsu	Mamberamo
Aquatic insects:		17 Heteroptera
Freshwater Fish:	1	
Frogs:	2	7
Reptiles:	Possibly 1	Possibly 1–3 species of lizards

Recommendations and Conservation Activities

Forests around Yongsu appear to be relatively secure due to local community support for conservation initiatives in the area. Ensuring continued preservation of forests at Yongsu will require further collaboration with these communities. Given its proximity to Jayapura and its excellent forest, Yongsu is an eminently suitable locality to continue the recently established training program for Papuan scientists. By employing and training local people at the center, which depends on the presence of intact habitats, it is hoped that local communities will continue to reap benefits from the preservation of their forests.

The RAP expedition to the Mamberamo Basin confirmed that the flora and fauna of this wilderness area are exceptionally diverse but still very poorly documented. Additional biodiversity surveys at a range of altitudes are still required to determine the status and distribution of rare and threatened species and to more comprehensively document patterns of species diversity and endemism within the basin.

Large projects, such as the Mamberamo Mega-Project, that will substantially alter and damage the forest and aquatic ecosystems should be resisted and local communities should be encouraged instead to pursue ecologically sound development projects.

The low human population density and considerable extent of remaining forest make the Mamberamo Basin the ideal site for biodiversity conservation. It is one of the largest tracts of pristine rainforest left in the world. These RAP results should be used by CI-Papua, local communities, and others to revise and strengthen the existing protected area system and to determine key areas for biodiversity conservation and community resource use within the basin.

Laporan Ringkas

PENILAIAN KONDISI BIOLOGI DI YONGSU-PEGUNUNGAN CYCLOPS DAN SELATAN SUNGAI MAMBERAMO, PAPUA, INDONESIA

Tanggal Ekspedisi
Pelatihan RAP di Yongsu: 19–30 Agustus 2000
Ekspedisi RAP Mamberamo: 1–15 September 2000

Deskripsi Lokasi
Pelatihan RAP dilaksanakan di sekitar Yongsu Dosoyo, sebelah utara Cagar Alam Pegunungan Cyclops, Propinsi Papua, Indonesia. Kawasan ini memiliki kontur topografi yang tinggi dengan punggung gunung yang berlekuk-lekuk dengan ketinggian lebih dari 1000 m hingga tepi laut dengan jarak kurang dari 5 km. Tidak terdapat dataran pesisir dan pelatihan dilakukan pada hutan yang terletak antara ujung utara lereng Pegunungan Cyclops dan laut.

Ekspedisi RAP mengeksplore berbagai habitat darat dan air di dua lokasi sekitar kampung Dabra, sebelah selatan daerah aliran Sungai Mamberamo. Aliran sungai ini mendukung hutan hujan dataran rendah asli yang luas di sebelah utara dari Cordillera tengah. Kawasan ini merupakan daerah tangkapan air yang terluas, menampung semua aliran sungai-sungai kecil ke utara yang turun mulai dari pegunungan tengah antara perbatasan Papua New Guinea dan sekitar 137° bujur barat. Batas permukaan air di sungai utama sangat berfluktuasi sepanjang tahun, menciptakan berbagai habitat termasuk rawa-rawa dan hutan terendam, rawa berumput, sungai mati, dan danau kecil. Terdapat daerah peralihan dari dataran rendah ke hutan di kaki bukit di sebelah selatan Mamberamo dimana Cordillera tengah menjulang tajam dari dataran rendah berawa. Survey kami dilakukan pada musim kering ketika permukaan air di sungai utama dan aliran sungai kecil dari cordillera tengah secara relatif sedang rendah.

Alasan Pelaksanaan Pelatihan dan Survei RAP
Pelatihan RAP dilakukan karena di Papua belum banyak ilmuwan yang memiliki kemampuan untuk secara cepat mengumpulkan, menganalisa, dan menyebarluaskan informasi keanekaragaman hayati yang sangat penting untuk membuat rekomendasi konservasi yang memadai. Pelatihan ini juga memberikan kesempatan untuk melakukan survei flora dan fauna di pesisir utara Pegunungan Cyclops.

Daerah aliran Sungai Mamberamo menjadi fokus ekspedisi RAP karena luasnya daerah yang masih alami, hutan yang jarang penduduknya dan belum banyak didokumentasi serta menghadapi ancaman yangmeningkat. Rencana Megaproyek Mamberamo termasuk peleburan aluminium dan perluasan industri pertanian yang listriknya disuplai dari dam pembangkit listrik tenaga air, akan merendam sejumlah besar areal hutan. Dam juga akan mengubah pola aliran air di hilir dengan dampak serius pada dinamika ekologi di ekosistem hutan dan perairan. Migrasi manusia yang berhubungan dengan proyek tersebut tidak diragukan lagi akan memberi dampak budaya yang serius bagi masyarakat adat yang tinggal di daerah aliran Sungai Mamberamo. Jika proyek ini dilaksanakan maka akan menimbulkan bencana besar yang saat ini sudah memperoleh tekanan dari aktivitas penebangan hutan. Pada kondisi tersebut, kajian mengenai kondisi dan nilai keragaman hayati dari eksositem hutan rimba yang luas ini sangat diperlukan untuk membantu pengembangan strategi konservasi dan pembangunan yang tepat.

Hasil-hasil Utama

Dua puluh tiga ilmuwan lokal dari Universitas Cenderawasih Papua, LSM, dan lembaga-lembaga di bawah Departemen Kehutanan dilatih oleh 10 ilmuwan dalam dan luar negeri yang ahli untuk flora atau fauna di kawasan ini. Hutan di Yongsu merupakan sumberdaya penting bagi penduduk setempat tetapi tingkat pemanfaatan tumbuhan dan hewan tampak berkelanjutan dan masih terdapat hutan luas tak terganggu di sekitar kampung. Hal ini berbeda dengan daerah di sebelah timur-laut Pegunungan Cyclops yang hutannya sudah terdegradasi berat. Masyarakat Yongsu mendukung kegiatan konservasi di daerah ini sehingga untuk jangka panjang, kondisi hutan yang bagus di dearah Yongsu dapat dipertahankan.

Survei di Mamberamo bukti membuktikan bahwa hutan, ekosistem perairan, dan tanah di sungai pada kondisi yang bagus. Tingkat populasi penduduk masih tergolong sangat rendah dan total areal yang telah dikonversi menjadi kebun masih kecil. Namun, pengamatan udara tampak jelas adanya perluasan jalan logging ke arah dari sungai dari utara menuju Mamberamo. Berdasarkan jumlah spesies yang berhasil ditemukan selama survei, proporsi ikan introduksi ada pada tingkat yang mengkhawatirkan (17,1%), dan dampaknya pada ekosistem perairan serta spesies ikan asli sangat perlu untuk dikaji. Terlepas dari posisinya yang berdekatan, kedua lokasi utama survei (Furu dan Tiri) memiliki flora dan fauna yang agak berbeda. Terdapat lebih banyak vegetasi sekunder di Furu daripada di Tiri, akibat dari besarnya dampak kegiatan manusia di lokasi tersebut. Selain itu, adanya hutan bukit di Furu memungkinkan diperolehnya berbagai tumbuhan dan satwa yang berbeda. Diperlukan survey tambahan di kawasan ini pada beberapa rentang ketinggian untuk memperoleh kajian lengkap mengenai nilai keanekaragaman hayatinya.

Jumlah Spesies (ns = tidak disurvei)

	Yongsu	Mamberamo
Tumbuhan:	178	Tiri: 131
		Furu: 234
Serangga air:	ns	56
Kupu-kupu:	69	129
Ikan air tawar:	33	23
Ikan karang:	195	ns
Katak:	8	21
Reptilia:	26	36
Burung:	90	143
Mamalia:	ns	7

Spesies Baru yang Ditemukan

	Yongsu	Mamberamo
Serangga air:	17 Heteroptera	
Ikan air tawar:	1	
Katak:	2	7
Reptilia:	Kemungkinan 1	Mungkin antara 1 dan 3 spesies kadal

Rekomendasi dan Aktivitas Konservasi

Hutan di sekitar Yongsu tampaknya relatif terlindungi karena penduduk setempat mendukung inisiatif konservasi di daerah ini. Untuk memastikan berlanjutnya perlindungan hutan di Yongsu diperlukan kerjasama lebih lanjut dengan masyarakat di Yongsu. Karena letaknya yang dekat dari Jayapura dan hutannya yang sangat bagus, Yongsu merupakan tempat yang cocok untuk melanjutkan program pelatihan bagi ilmuwan Papua. Hal ini akan memungkinkan masyarakat lokal dapat terus memperoleh manfaat dari perlindungan hutan mereka.

Ekspedisi RAP ke Daerah Aliran Sungai Mamberamo memastikan bahwa flora dan fauna di rimba belantara ini adalah luar biasa beragam tetapi dokumentasinya masih sedikit. Diperlukan beberapa survei tambahan pada sejumlah ketinggian untuk menentukan status dan distribusi spesies yang terancam dan langka untuk melengkapi informasi mengenai pola keragaman spesies dan endemisitas di Mamberamo.

Proyek-proyek besar seperti Proyek Mega Mamberamo yang akan mengubah dan merusak eksosistem hutan dan perairan harus ditolak dan masyarakat lokal perlu didorong untuk mengembangkan proyek-proyek pembangunan yang memperhatikan aspek ekologi.

Kepadatan penduduk yang rendah dan hutan yang luas membuat Daerah Aliran Sungai Mamberamo menjadi tempat ideal untuk konservasi keanekaragaman hayati. Mamberamo merupakan salah satu hutan hujan asli terluas yang tersisa di dunia. Hasil RAP ini dapat digunakan oleh CI-Papua, masyarakat lokal, dan lembaga lainnya untuk merevisi dan memperkuat sistem kawasan lindung yang telah ada; dan menentukan daerah kunci di Mamberamo untuk konservasi keanekaragaman hayati dan pemanfaatan sumberdayanya oleh masyarakat.

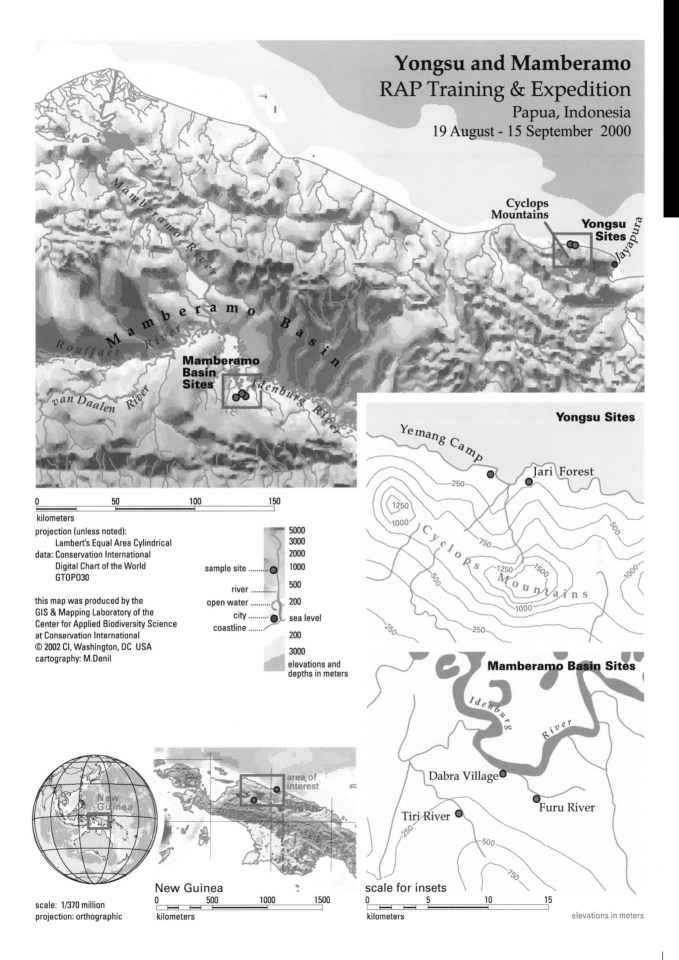

Yongsu and Mamberamo
RAP Training & Expedition
Papua, Indonesia
19 August - 15 September 2000

Cyclops Mountains

Yongsu Sites

Jayapura

Mamberamo River

Mamberamo Basin

Rouffaer River

van Daalen River

Idenburg River

Mamberamo Basin Sites

0   50   100   150

kilometers

projection (unless noted):
    Lambert's Equal Area Cylindrical
data: Conservation International
    Digital Chart of the World
    GTOPO30

this map was produced by the
GIS & Mapping Laboratory of the
Center for Applied Biodiversity Science
at Conservation International
© 2002 CI, Washington, DC USA
cartography: M.Denil

5000
3000
2000
sample site 1000
500
river ....
open water 200
city .... sea level
coastline 200
3000

elevations and
depths in meters

Yongsu Sites

Yemang Camp

Jari Forest

250

1250

1000

750

Cyclops

500

1250

1500

Mountains

1000

1000

250

250

500

Mamberamo Basin Sites

Idenburg River

Dabra Village

Furu River

Tiri River

250

500

750

New Guinea

area of interest

scale: 1/370 million
projection: orthographic

New Guinea
0   500   1000   1500

kilometers

scale for insets
0   5   10   15

kilometers

elevations in meters

YONGSU

Rainforest extends to the coast at Jari near Yemang Camp.

Stephen J. Richards

Hook-billed Kingfisher (*Melidora macrorrhina*) from forest at Yongsu.

Stephen J. Richards

Gerald R. Allen

Lentipes multiradiatus, a new species of freshwater goby discovered during the Yongsu training survey.

RAP trainees and scientists taking a GPS reading at Yongsu.

Debbie Gowensmith

RAP leaders Suer Suryadi, Stephen Richards, and Bruce Beehler with guide Pak Simon (second from left) at Yongsu.

Debbie Gowensmith

MAMBERAMO RAP SURVEY

Stephen J. Richards

Lowland and foothill rainforest around Furu Camp, Mamberamo Basin.

Stephen J. Richards

Most of the RAP Team members at Tiri Camp.

Stephen J. Richards

Forest interior near Dabra Village.

Stephen J. Richards

This small feather-tailed possum (*Distoechurus pennatus*) was found in the forest around Furu Camp.

Stephen J. Richards

D'Albertis Python (*Leiopython albertisi*) found in forest near Dabra Village.

Gerald R. Allen

Glossamia beauforti (Beaufort's Mouth Almighty), known from northern Papua.

Stephen J. Richards

A bizarre microhylid frog (*Asterophrys turpicola*) from the forest floor around Tiri camp.

Stephen J. Richards

Botanist Ismail Rachman examines a specimen at Tiri Camp.

Stephen J. Richards

An undescribed treefrog (*Litoria* sp.) from forest near Tiri Camp.

Stephen J. Richards

A giant soft-shelled turtle (*Pelochelys cantori*) from Tiri River. A favored food item for local hunters.

Executive Summary

INTRODUCTION

Papua (previously known as Irian Jaya) is Indonesia's easternmost and least-developed province. It incorporates the western part of New Guinea, the world's largest and highest tropical island, and a number of smaller islands nearby. Though biological data for Papua is sorely lacking, it has been estimated that the province contains around fifty percent of Indonesia's total biodiversity. Exceptionally high levels of diversity and endemism among the flora and fauna (Conservation International, 1999) reflect New Guinea's long isolation from other land masses and an extraordinary array of ecosystems ranging from coastal savannas and tropical forests to alpine habitats on the highest mountains in the Asia-Pacific region. Papua contributes significantly to Indonesia's status as one of the biologically richest countries in the world. In recognition of its high biodiversity values and low human population density Conservation International (CI) in 1997 declared New Guinea a "*Major Tropical Wilderness Area*" (TWA). With at least 70% of original forest cover intact, TWAs are important storehouses of biodiversity and act as major watersheds. They are also places where indigenous people have the opportunity to maintain their traditional lifestyles.

Papua's forests have until recently escaped the devastating extent of logging typical of Indonesia's western provinces. However, between 1993 and 1997, forest cover in Papua decreased from ninety percent to eighty percent (Conservation International, 1999) and rates of forest conversion in Papua have almost certainly increased during Indonesia's recent economic crisis. Loss of forest due to logging, transmigration, plantation agriculture, mining, oil and gas operations, and infrastructure projects associated with human settlement and roads threatens to destroy much of Papua's unique biological heritage. Increasing pressure for development means that people responsible for managing and conserving Papua's natural resources face a tremendous need for biological information.

Recognising that conservation in Papua is constrained by a lack of basic information about the diversity, distribution, and abundance of flora and fauna, the Irian Jaya Conservation Priorities Setting Workshop in 1997 concluded that capacity building of local scientists and the collection of biological information are urgently required to ensure that informed conservation recommendations can be formulated for the province. CI has joined forces with provincial authorities, local universities and other institutions, and local communities and NGOs to achieve these aims. In March 1998, CI and the Indonesian Institute of Sciences (LIPI) undertook the first Rapid Assessment Program (RAP) survey in Papua (Mack and Alonso, 2000). The Wapoga River survey documented a large number of plants and animals previously unknown to science, underlining the fact that to make informed policy decisions about development and conservation in Papua will require collection of substantially more baseline biological data than are currently available.

One reason why so much of Papua remains biologically unknown is the lack of skills and capacity to conduct biodiversity surveys. To improve local capacity to undertake scientific studies that will rapidly accumulate biological data for use by decision-makers, CI-Indonesia conducted training courses in 1999–2000 to train Indonesian biologists in the field skills required for rapid biological surveys, and in research design and data analysis and presentation. These training courses are a long-term investment by CI to produce a team of locally based RAP scientists with skills in identification of specific taxa groups and the ability to respond rapidly to the need for biodiversity information anywhere in the province.

As part of this program CI and the University of Cenderawasih (UNCEN), in collaboration with local conservation NGOs, undertook a combined training and biological survey project in Papua from 19 August until 15 September 2000. Twenty-three participants, many from UNCEN and local NGOs, joined scientists at Yemang research camp on the northern edge of the Cyclops Mountains near the village of Yongsu Dosoyo (02°25.994'S, 140°29.147'E) from 19–30 August for a hands-on training course. The village is on the edge of the Cyclops Mountains Nature Reserve, and is approximately 30 km west of Jayapura, the provincial capital of Papua. Training and surveys at Yongsu were conducted in the vicinity of the training camp and in nearby Jari forest, a short boat-ride from camp.

Six participants, each focusing on one taxonomic group, were chosen to work alongside scientists in an intensive follow-up training survey near the village of Dabra in the Mamberamo River Basin from 1–15 September 2000. The vast and near-pristine wilderness of the Mamberamo River Basin was selected for the training expedition because it was identified during the Irian Jaya Biodiversity Conservation Priority-Setting Workshop in early 1997 as an area of probable biological significance where surveys are urgently required to document levels of biodiversity and endemism. The expedition involved scientists from the United States (2), Australia (2), the Netherlands (1), Papua New Guinea (1), and Indonesia (4). These scientists provided participants with further training in rapid-inventory techniques at two sites in the Mamberamo basin: Furu River (3°17'04"S, 138°38'10"E) and Tiri River (3°17'30"S, 138°34'53"E). Both sites are in lowland rainforest (between 80 m and 90 m asl) within a 5 km radius of Dabra village. South of the Furu camp the forest is in a transition zone between flat, swampy forest typical of the Mamberamo River Basin and hill forest at the base of the steep northern slopes of New Guinea's central cordillera. Though hunting pressure, swidden agriculture, and selective logging affect some forest areas, the human population density of the region is very low. Six taxonomic groups were inventoried: birds, freshwater fish, herpetofauna, insects (specifically aquatic insects and butterflies), small mammals, and vegetation.

SUMMARY OF RESULTS

The training program is an integral part of CI's long-term strategy to build local capacity so that local scientists can participate in future RAP activities and conduct independent biological research and surveys. The RAP training course was attended by 23 trainees from local universities (UNCEN and UNIPA), Forestry Department (National Parks, Nature Conservation Agencies, and Forestry Research & Development Institute), and NGOs. Scientists with extensive research experience in the region demonstrated the latest RAP survey techniques, provided trainees with instruction on scientific methods and field research techniques, data analysis and presentation, and scientific writing, and held discussions on biodiversity and conservation issues in Papua and PNG.

Vegetation: Yongsu

Plant diversity and forest composition in lowland rainforest at Yongsu, Papua, were studied using five 0.1 ha modified Whittaker plots and general surveys. One hundred and ninety-seven plant species were recorded. The canopy is dominated by *Manilkara fasciculata, Mastixiodendron pachyclados, Palaquium ridleyi,* and *Parastemon urophyllus.* Forests around Yongsu provide valuable material for house and canoe construction by local villagers. However, vegetation in the Yongsu area, especially at Jari, remains largely intact and extraction of forest resources appears to be sustainable at this stage. Floral richness in lowland forest at Yongsu is similar to other tropical regions in the Asia-Pacific region but is lower than in Latin America.

Vegetation: Mamberamo

This study documented the composition and diversity of vegetation in 0.5 ha at two sites in the Dabra area, using line transects (Tiri), modified Whittaker plots (Furu), and general surveys (both sites). A total of 234 plant species were recorded at Furu and 131 species were documented at Tiri, but species accumulation curves indicate that the flora was not comprehensively sampled. There is substantial overlap in composition of the 10 most abundant species at the two Mamberamo sites. Exceptions are *Pometia pinnata* that is abundant at Tiri but not at Furu, *Vatica rassak* that is abundant at Furu but not at Tiri, and *Ceratopetalum succirubun,* that is abundant at Furu but absent from Tiri. The abundance of *Pometia pinnata* at Tiri can be attributed to the greater disturbance of forest at that site. The poor state of knowledge of the taxonomy and distribution of Mamberamo vegetation precluded assessment of the conservation status of almost all species recorded in this study. Many of the botanical specimens were previously not represented in the Herbarium Bogoriense, Bogor.

Aquatic insects: Mamberamo

Aquatic insects were sampled at nine stations in the Dabra area of the Mamberamo River Basin between 5 and 15 September 2000 by visual searching, hand netting, and localized pyrethrin fogging of riparian logs and hygropetric habitats. At least 40 species in 23 genera of aquatic Heteroptera (true bugs) were collected, of which at least 17 are new to science. Fourteen species in nine genera of Zygoptera (damselflies), all previously described, were collected. Two species in two genera of Gyrinidae (whirlygig beetles) were collected, with the number of new species, if any, uncertain pending detailed analysis. Aquatic insect diversity in the Dabra area is high, and similar to that seen at corresponding elevational zones in the Wapoga River Basin, which lies immediately to the west of the Mamberamo.

In the garden areas northwest of Dabra, streambeds have been affected by the loss of forest cover on the surrounding slopes. Additional clearing of forest will cause erosion into streams, degrading the water quality and affecting the system's diversity. The river system, however, seems healthy overall; during the survey, a swarm of mayflies hatched, indicating that toxic chemicals are not present in river-bottom soils.

Butterflies: Yongsu

Sixty-nine species of butterflies in four families were recorded using long-handled nets and fluorescent and mercury vapour lights. The fauna comprised: Papilionidae (7 species), Pieridae (7 species), Lycaenidae (17 species), and Nymphalidae (38 species). The nymphalid *Elymnias paradoxa* is a species previously known only from eastern Papua New Guinea, and the record of this species at Yongsu is a significant westerly range extension. Total diversity at this lowland site represents nearly half the 144 species reported from the entire Cyclops Mountains Nature Reserve, suggesting that the forests at Yongsu provide good-quality habitats for these insects. Species accumulation curves indicate that additional species are likely to occur in this area.

Butterflies and moths: Mamberamo

One hundred and twenty-nine species of butterflies in 68 genera were collected using long-handled nets. These include representatives of all nine families of butterflies known from Papua. The butterfly fauna is dominated by species typical of Papua's lowland coastal areas. More than 480 species of moths in over 112 genera were collected from large white screens illuminated by 160 Watt white lamps. Moths of the family *Pyralidae* were particularly diverse with over 145 species, many of which remain unidentified. Large moths of the families *Sphingidae* (only 5 species) and *Saturniidae* (none) were very poorly represented. The overall diversity and abundance of the lepidopteran fauna in the Mamberamo area indicates that the forest is still in good condition. Further studies are required to determine the full diversity of the butterfly fauna in this region, especially at higher elevations, and to assess the status of rare and protected species.

Fishes: Yongsu

The freshwater fish fauna was documented primarily by visual surveys with a mask and snorkel but seine nets of various sizes, hand nets and ichthyocide rotenone application were used occasionally. The fauna consists of 33 species in 25 genera and 15 families. Yongsu fishes are adapted to relatively steep-gradient streams, and the fauna is similar to other assemblages inhabiting mountainous coastlines on the north coast of New Guinea. Gobioid fishes (Gobiidae and Eleotridae) dominate the fauna, accounting for nearly half of the total fishes. The cling-goby subfamily Sicydiinae is particularly well represented with seven species. The Yongsu area and adjacent Cyclops coast may provide the best example of steep-gradient coastal stream habitat on the entire mainland of Papua. It is also the home of two apparently endemic gobies in the genus *Lentipes*, one of which is a new species discovered during this training course, thus providing justification for conservation initiatives.

Fishes: Mamberamo

A total of 23 species in 18 genera and 11 families were documented using nets of various sizes, visual surveys, application of ichthyocide rotenone, and examination of locally caught fish in the Dabra market. The small goby *Gobius tigrellus* was collected for the first time since the 1930s. It was previously known from only 10 specimens collected from the Mamberamo system by the 1938–1939 Archbold Expedition. Unfortunately, six introduced fish species were also documented in the Mamberamo system, which has the highest proportion of endemic species of any New Guinea river. The impacts of these introduced species on native fish populations should be assessed as a matter of urgency.

Amphibians and reptiles: Yongsu

Twenty-six species of reptiles and eight species of frogs were recorded from the Yongsu area using a combination of visual and aural surveys and litter plots. Two species of frogs are new to science and another frog heard calling from high in the forest canopy almost certainly represents an undescribed species of *Litoria*. Further attempts should be made to collect and identify this potentially very interesting canopy-dwelling species.

Five species (2 frogs, 3 reptiles) were recorded in litter plots. A species accumulation curve for litter herpetofauna reached an asymptote after just 8 plots but general searching of the forest floor demonstrated that the litter herpetofauna is substantially more diverse than indicated by these plots. Considerable time and effort are required to establish and search plots, and this technique did not reveal any species that were not detected during general surveys. Litter plots are not particularly useful for rapid, comprehensive inventories of New Guinea litter herpetofauna.

Two species of marine turtle nest on the beaches at Yongsu: the green turtle (*Chelonia mydas*) and the loggerhead turtle (*Caretta caretta*). Green sea turtle hatchlings emerged from a nest at Yongsu camp at 6:30 p.m. on 25 August 2000 and a single loggerhead turtle hatchling was observed at the same locality on 28 August 2000.

Amphibians and reptiles: Mamberamo

Twenty-one species of frogs and 36 species of reptiles were recorded from three sites in the Mamberamo River Basin (Furu, Tiri, and the immediate vicinity of Dabra village) using visual surveys during the day and visual and aural surveys at night. Turtles were surveyed using a mask and snorkel and were also obtained from local hunters. Seven species of frogs and up to three species of lizards are unknown to science. Documentation of the Sail-fin Lizard *Hydrosaurus amboinensis* at Furu and Tiri represents a significant easterly range extension for this large and spectacular reptile. Several large reptiles including the Freshwater Crocodile (*Crocodylus novaeguineae*) and two freshwater turtles—the Softshell Turtle (*Pelochelys cantori*) and the New Guinea Side-Neck (*Elseya novaeguineae*)—are harvested by local communities and represent a source of cash income and food. Harvesting of crocodiles in the area appears to have been unsustainable and local villagers complained that crocodile populations have dropped to alarmingly low numbers. Sustainability of turtle harvesting should also be assessed as a matter of priority.

Birds: Yongsu

Birds were documented in a tract of selectively logged hill forest on a dissected and sloping narrow plateau (40–70 m asl) about 750 m south of the Yemang Training Camp. The survey was conducted over a ten-day period (20–29 August 2000) using mist-netting, audial censuses, point counts, and *ad lib* observations. Point counts indicated that the forest avifauna is quite patchy at a local level, possibly in response to distribution of food resources; informal censuses of forest trees suggested that the arboreal forest flora is patchily distributed. Ninety species of birds were recorded, and most of these were forest-dwelling birds. Fifteen birds typical of lowland forest, including the Victoria Crowned Pigeon and the Common Paradise-Kingfisher were apparently absent. Conversely, the presence of healthy populations of Blyth's Hornbill, Palm Cockatoo, and Northern Cassowary indicate that this forest is not depleted of its megafauna. Lack of a coastal plain, and lack of significant lowland forest are factors that probably contribute to the relatively species-poor bird fauna. There was no evidence that small-scale selective logging in this forest has harmed local bird populations.

Reports from a local naturalist provided evidence that the Cyclops Mountains remain significantly under-surveyed for birds and mammals.

Birds: Dabra

A total of 143 species of birds were recorded from lowland forest, swamps, and riverine habitats in the Dabra region using mist-nets and visual and audial transects. Of these 65 (45%) are endemic to New Guinea. The bird species sighted included six Birds of Paradise (the Lesser, the King, the Twelve-Wired, the Jobi Manucode, the Glossy-Mantled Manucode, and the Riflebird), bowerbirds, and catbirds. Globally threatened species that are vulnerable to hunting, such as Northern Cassowary and Victoria Crowned Pigeon, were relatively common in the area, indicating that hunting pressure from local communities is low. Rare local endemics such as the Pale-billed Sicklebill and Brass's Friarbird were not found during the present survey, but the extensive swamp forests to the north of Dabra appear to be suitable habitat for those two species and should be surveyed in the near future.

Small mammals: Dabra

Small mammals were trapped in primary forest around the Furu River and Tiri River sites using 71 Elliot traps (497 trap nights) and 5 mist nets (35 trap nights). Sixty-nine mammals representing seven species were trapped during the survey. In total four species of bats, two species of rodents, and one species of marsupial were positively identified. Fauna at the two sites was similar; the same four bat species were collected at each site, but no rodents or marsupials were collected at the Tiri River site. All of the bats are megachiropteran fruit bats in the family Pteropodidae.

One female *Syconycteris australis* at Furu was carrying an embryo, and one *S. australis* and one *Paranyctimene raptor* were lactating. Eight *Nyctimene draconilla* were also carrying young embryos.

Collection of the feather-tailed possum *Distoechurus pennatus* at Furu Camp fills a wide gap in its known distribution and is one of few records for this species in Papua (Flannery 1995). The White-bellied Melomys (*Melomys leucogaster*) was previously known only from southern New Guinea (Flannery 1995), and this is the first record from north of the central cordillera. Survey effort at the Dabra sites was too limited to produce a comprehensive inventory. However, documentation of two significant distributional records indicate that additional surveys are likely to produce more exciting discoveries in this poorly known area.

OVERALL SUMMARY AND CONSERVATION RECOMMENDATIONS

The training component of the 2000 Papua RAP program substantially improved local capacity for rapid biodiversity inventory studies. The data accumulated during the training surveys also highlighted the biological significance of forests in the Yongsu and Dabra areas. In particular the discovery of new species and the large range extensions for a number of other species reinforce how poorly known these regions are. The fauna and flora of the Mamberamo Basin in particular require substantial additional documentation. Carefully targeted surveys at a range of altitudes in other regions of the Mamberamo Basin should be undertaken to provide a broader and more accurate assessment of the biodiversity and conservation significance of this vast wilderness area. Understanding the ecological processes and linkages within and among different sites in the basin will be critical for formulating the most appropriate conservation strategies for the region.

Though lightly populated and retaining most of its forest cover, the Mamberamo Basin faces significant threats. The proposed Mamberamo Mega-Project, which includes a hydro-electric dam, would have serious deleterious effects on the human population and biodiversity of the Mamberamo

River basin. Further confirmation of the region's suspected significant biodiversity values is therefore a high priority so that accurate information can be provided to decision-makers involved with plans for major development projects in the basin. Hunting pressure may already be affecting populations of cassowaries, turtles, and particularly crocodiles, and these pressures will increase with increasing human population and the widespread availability of guns. Current hunting practices and target species should be documented so that potential impacts can be assessed and sustainable harvesting rates can be promoted.

The following conservation and research recommendations are developed from the chapters in this report. Many will require cooperation between local communities, NGOs, and the provincial authorities. Addressing the issue of exotic species introductions will also require education of and collaboration with authorities elsewhere in western Indonesia.

Yongsu

- Provide further opportunities and technical assistance for local biologists to develop their scientific skills through additional training courses in forests of the Yongsu area. These courses should be designed to maximize the collaboration of, and benefits for, local landowners.

- Monitor turtle hunting and excavation of turtle nests to determine sustainability of current harvesting practices.

- Placement and design of roads into and around the periphery of the Cyclops Mountains should take into account potential impacts on local flora and fauna.

- Undertake additional higher-altitude biodiversity surveys in the Cyclops Mountains.

Mamberamo

- Enhance conservation prospects in the Mamberamo River Basin by undertaking a campaign in the international and national community to promote the Mamberamo Basin as the last great wilderness frontier for Papua.

- Promote conservation awareness about the region through a two-pronged education program. One approach should be aimed at the provincial and national governments and involve development of curricula outlining the significance of the biodiversity, resources, and cultures of the Mamberamo basin. The second approach should focus at the community level and develop educational programs to engender local pride in the biological and cultural values of the region. This program should be aimed at local school children and adults. Education at the community level should also promote skills that

will add value to existing resource extraction activities. Local communities will initially require technical assistance for resource evaluation, resource mapping, and skill assessment.

- Economic analyses of transport costs and requirements, and costs of market development and access should be undertaken for prospective projects in the basin. These analyses will be critical for promoting economically and ecologically robust projects with minimal destruction of resources and biodiversity.

- The population status and harvesting levels of crocodiles and other large aquatic reptiles should be assessed to ensure long-term sustainability of these resources.

- Opportunities and assistance for local scientists to undertake biological, social, and cultural studies in the basin should be provided.

- Further surveys should be undertaken in the basin to more comprehensively document the biodiversity values of this region and to assess the status of rare and threatened species. The highest priority areas for future RAPs are the Foja Wildlife Sanctuary and the Roffaer Wildlife Sanctuary.

- Exotic (non-native) species pose a serious threat to native forests and waterways in Papua. Unfortunately, these species are virtually impossible to eradicate and future conservation measures should include education programs for government officials and others involved in the transportation of introduced species to explain the need to prevent future introductions of alien species. The effects of introduced fish species on native fish populations in the Mamberamo River system should be documented and this information used as part of the education program to prevent further introductions into this and other aquatic systems in Papua.

LITERATURE CITED

Conservation International. 1999. The Irian Jaya biodiversity conservation priority-setting workshop. Final Report. Washington, DC: Conservation International.

Flannery, T.F. 1995. Mammals of New Guinea (2nd ed.). Chatswood: Reed Books.

Mack, A.L. and Alonso, L.E. (eds.). 2000. A biological assessment of the Wapoga River area of northwestern Irian Jaya, Indonesia. RAP Bulletin of Biological Assessment Number 14. Washington, DC: Conservation International.

Ringkasan Eksekutif

PENDAHULUAN

Papua (sebelumnya dikenal dengan Irian Jaya) adalah propinsi Indonesia paling timur dan paling kurang berkembang. Terdiri dari sebelah barat New Guinea, pulau tropis tertinggi dan terluas di dunia, dan sejumlah pulau-pulau kecil di sekitarnya. Walaupun data biologi untuk Papua belum banyak diketahui, propinsi ini diduga memiliki sekitar lima puluh persen total keanekaragaman hayati Indonesia. Tingginya tingkat keanekaragaman dan endemisitas flora dan fauna yang luar biasa (Conservation International, 1999) menunjukkan lamanya New Guinea terisolasi dari daratan utama lainnya dan menunjukkan adanya berbagai tipe ekosistem mulai dari pesisir dan hutan tropis hingga habitat alpin di pegunungan tertinggi di kawasan Asia Pasifik. Papua memberikan sumbangan besar pada status Indonesia sebagai salah satu Negara terkaya di dunia dari aspek biologi. Karena tingginya nilai keragaman hayati dan rendahnya kepadatan populasi manusia maka Conservation International (CI) pada 1997 menyatakan New Guinea sebagai "*Major Tropical Wilderness Area*" (TWA) atau Kawasan Rimba Tropis Utama. Dengan paling sedikit 75% tutupan hutannya masih alami, maka TWA merupakn gudang keragaman hayati yang penting, dan berperan dalam tata air. TWA juga merupakan tempat bagi masyarakat asli memiliki kesempatan untuk memelihara pola hidup mereka.

Hutan-hutan di Papua hingga kini belum mengalami kehancuran sebagaimana aktivitas pembalakan kayu di wilayah barat Indonesia. Namun demikian, antara 1993 dan 1997 tutupan hutan di Papua menurun dari sembilan puluh persen menjadi delapan puluh persen (Conservation International, 1999) dan tingkat konversi hutan di Papua meningkat ketika krisis ekonomi menimpa Indonesia. Berkurangnya hutan akibat penebangan, transmigrasi, perkebunan, pertanian, pertambangan, penambangan minyak dan gas, serta proyek infrastruktur yang berkaitan dengan perumahan dan jalan merupakan ancaman bagi kerusakan sebagian besar warisan biologi Papua yang unik. Meningkatnya tekanan pembangunan mengakibatkan orang yang bertanggung jawab mengelola dan mengkonservasi sumber daya alam Papua sangat memerlukan informasi biologi.

Menyadari bahwa konservasi di Papua terhambat oleh kurangnya informasi dasar tentang keragaman, distribusi, dan kelimpahan flora dan fauna, maka Lokakarya Penentuan Prioritas Konservasi Irian Jaya pada tahun 1997 menyimpulkan bahwa pengembangan kapasitas ilmuwan lokal dan pengumpulan informasi biologi adalah sangat diperlukan untuk memastikan agar-dabat bagi rekomendasi konservasi Jaya merumuskan propinsi ini. CI bekerjasama dengan pemerintah propinsi, universitas lokal dan institusi lainnya, serta masyarakat lokal dan LSM berupaya untuk mewujudkan tujuan tersebut. Pada bulan Maret 1998, CI dan Lembaga Ilmu Pengetahuan Indonesia (LIPI) melakukan Rapid Assessment Program (RAP) yang pertama di Papua (Mack dan Alonso, 2000). Survei di Sungai Wapoga berhasil menemukan banyak tumbuhan dan satwa yang belum pernah diketahui oleh ilmu pengetahuan, membuktikan bahwa untuk membuat kebijakan pembangunan dan konservasi yang memadai di Papua akan membutuhkan lebih banyak lagi data dasar biologi daripada yang telah tersedia sekarang ini.

Salah satu alasan mengapa begitu banyak informasi biologi belum diketahui adalah kurangnya kapasitas dan keahlian untuk melakukan survei keanekaragaman hayati. Untuk meningkatkan kapasitas lokal melakukan kajian ilmiah yang secara cepat dapat mengakumulasi data biologi untuk dimanfaatkan oleh para pengambil keputusan, CI-Indonesia melaksanakan program pelatihan pada tahun 1999–2000 untuk melatih ahli biologi Indonesia agar mampu melakukan survei biologi dengan cepat, merancang penelitian, serta analisa data dan presentasi. Program pelatihan ini merupakan upaya jangka panjang CI untuk menghasilkan tim RAP lokal dengan keahlian mengidentifikasi kelompok taksa tertentu dan kemampuan merespon dengan cepat kebutuhan informasi keragaman hayati di manapun di Papua.

Sebagai bagian dari program peningkatan kapasitas tersebut, CI dan Universitas Cenderawasih (UNCEN), bekerjasama dengan LSM konservasi lokal melakukan kegiatan pelatihan yang dilanjutkan dengan survei biologi sejak 19 Agustus hingga 15 September 2000. Dua puluh tiga peserta, kebanyakan berasal dari UNCEN dan LSM lokal, bergabung dengan para ilmuwan di Camp Penelitian Yemang di sebelah utara Pegunungan Cyclops dekat kampung Yongsu Dosoyo (02°25.994S, 140°29.147T) pada tanggal 19–30 Agustus untuk dilatih oleh ahlinya. Kampung Yongsu terletak dekat Cagar Alam Pegunungan Cyclops, sekitar 30 km sebelah barat Jayapura, ibu kota propinsi Papua. Pelatihan dan survei dilakukan di sekitar camp pelatihan dan sekitar hutan Jari, yang letaknya tidak jauh dari camp.

Enam peserta, yang mewakili tiap taksa dipilih untuk bersama-sama ilmuwan melakukan pendalaman dan penerapan hasil pelatihan di sekitar kampung Dabra, daerah aliran Sungai Mamberamo pada tanggal 1–15 September 2000. Rimba utama yang luas dan masih asli di daerah aliran Sungai Mamberamo dipilih sebagai tempat ekspedisi dalam pelatihan ini karena dalam Lokakarya Penentuan Kawasan Prioritas Konservasi Keanekaragaman Hayati Irian Jaya tahun 1997 dinyatakan sebagai daerah yang kemungkinan memiliki nilai biologi penting sehingga perlu segera disurvei untuk mendokumentasikan tingkat keragaman hayati dan endemisitasnya. Ekspedisi ini melibatkan ilmuwan dari Amerika Serikat (1), Australia (2), Belanda (2), Papua New Guinea (1) dan Indonesia (3). Para ilmuwan itu memberikan teknik-teknik tingkat lanjut untuk melakukan inventori secara cepat kepada para peserta di dua lokasi di Mamberamo, yaitu Sungai Furu (3°17'04"S, 138°38'10"T) dan Sungai Tiri (3°17'30"S, 138°34'53"T). Kedua lokasi terletak pada hutan hujan dataran rendah (antara 80 m dan 90 m dpl) dalam jarak radius 5 km dari Kampung Dabra. Hutan di sebelah Selatan camp Furu merupakan daerah transisi antara dataran, hutan rawa tipikal daerah aliran Sungai Mamberamo dan hutan bukit di kaki lereng utara Cordillera tengah New Guinea. Walaupun tekanan perburuan, kebun, dan penebangan selektif mempengaruhi beberapa kawasan hutan, kepadatan populasi penduduknya masih sangat rendah. Enam kelompok taksonomi yang diteliti: burung, ikan air tawar, herpetofauna, serangga (khususnya serangga air dan kupu-kupu), mamalia kecil, dan vegetasi.

RINGKASAN HASIL

Program pelatihan merupakan bagian dari strategi jangka panjang CI untuk membangun kapasitas lokal sehingga ilmuwan lokal dapat berpartisipasi dalam kegiatan RAP di masa mendatang, serta mampu melakukan survei dan penelitian biologi. Pelatihan RAP diikuti oleh dua puluh tiga ilmuwan dari universitas lokal (UNCEN dan UNIPA), Departemen Kehutanan (Taman Nasional, KSDA, dan Litbang Kehutanan), dan LSM. Para ilmuwan yang berpengalaman di kawasan ini memberikan berbagai teknik mutakhir untuk melakukan survei RAP, metodologi ilmiah, teknik-teknik penelitian lapangan, analisa data dan presentasi, penulisan ilmiah, dan melakukan diskusi mengenai keanekaragaman hayati dan isu-isu konservasi di Papua dan PNG.

Vegetasi: Yongsu

Komposisi hutan dan keragaman tumbuhan di hutan hujan dataran rendah Yongsu, Papua, diteliti dengan menggunakan 0,5 ha plot Whittaker yang telah dimodifikasi dan survei umum. Seratus tujuh puluh delapan spesies berhasil ditemukan. Tingkat kanopi didominasi oleh *Manilkara fasciculata, Mastixiodendron pachyclados, Palaquium ridleyi* dan *Parastemon urophyllus*. Hutan di sekitar Yongsu merupakan bahan penting untuk pembuatan rumah dan perahu oleh penduduk setempat. Namun demikian, vegetasi di daerah Yongsu, khususnya di Jari, masih dalam keadaan baik dan ekstraksi sumberdaya hutan pada saat ini tampaknya dalam kondisi lestari berkelanjutan. Kekayaan jenis tumbuhan di hutan dataran rendah Yongsu mirip dengan kawasan tropis lainnya di Asia Pasifik tetapi lebih rendah daripada Amerika Latin.

Vegetasi: Mamberamo

Penelitian ini mendokumentasikan komposisi dan keragaman vegetasi pada dua tempat di daerah Dabra, dengan menggunakan garis transek (Tiri), 0,5 ha plot Whittaker yang dimodifikasi (Furu), dan survei umum (kedua tempat). Total sebanyak 234 spesies tumbuhan berhasil ditemukan di Furu dan 131 spesies di Tiri, tetapi kurva akumulasi spesies menunjukkan bahwa tumbuh-tumbuhan belum disampling dengan lengkap. Terdapat tumpang tindih yang besar pada komposisi 10 spesies paling melimpah di dua lokasi tersebut. Terdapat pengecualian pada *Pometia pinnata* yang melimpah di Tiri tapi tidak melimpah di Furu, *Vatica rassak* melimpah di Furu tapi tidak melimpah di Tiri, dan *Ceratopetalum succirubun*, melimpah di Furu tetapi tidak ditemukan di Tiri. Kelimpahan *Pometia pinnata* di Tiri dapat disebabkan oleh besarnya gangguan pada hutan di lokasi tersebut. Rendahnya pengetahuan mengenai taksonomi dan distribusi vegetasi Mamberamo menjadi hambatan untuk menentukan status konservasi dari sebagian besar spesies yang ditemukan dalam penelitian ini. Banyak spesimen contoh tumbuhan yang dikumpulkan dari lokasi ini tidak terdapat di Herbarium Bogoriense, Bogor.

Serangga air: Mamberamo

Serangga air dikumpulkan dari sembilan stasiun di daerah Dabra, daerah aliran sungai Mamberamo antara 5 dan 15 September 2000 dengan pencarian visual, jaring tangan, dan pengabutan terbatas "pyrethrin" dari kayu-kayu di tepi sungai dan habitat "hygropetric." Sedikitnya berhasil ditemukan 40 spesies dari 23 genus Heteroptera air, yang 17 diantaranya merupakan spesies baru bagi ilmu pengetahuan. Selain itu berhasil pula ditemukan empat belas spesies dari sembilan genus Zygoptera, semuanya telah dideskripsikan. Dua spesies dari 2 genus Gyrinidae berhasil dikoleksi, dengan sejumlah spesies baru, jika ada, yang memerlukan analisa lebih lanjut. Keragaman serangga air di Dabra tergolong tinggi, dan mirip dengan yang telihat pada zona elevasi yang sama di daerah aliran Sungai Wapoga, yang letaknya di sebelah barat Mamberamo.

Kebun penduduk di sebelah barat laut Dabra telah dipengaruhi oleh hilangnya hutan di sekitar lereng. Bertambahnya penebangan hutan akan menimbulkan erosi ke aliran sungai, menurunnya kualitas air, dan mempengaruhi keragaman hayatinya. Namun demikian, secara keseluruhan perairan sungai tampaknya dalam keadaan sehat/baik; selama survei, adanya kerumunan "mayflies" yang baru menetas mengindikasikan bahwa tidak terdapat kimia beracun pada tanah di dasar sungai.

Kupu-kupu: Yongsu

Enam puluh sembilan spesies kupu-kupu dari dari 4 famili berhasil ditemukan dengan menggunakan jaring penangkap dan layar dengan lampu merkuri dan neon. Fauna yang berhasil dikumpulkan terdiri dari: Papilionidae (7 spesies), Pieridae (7 spesies), Lycaenidae (17 spesies) dan Nymphalidae (38 spesies). Kupu-kupu Nymphalidae, *Elymnias paradoxa* adalah spesies yang sebelumnya hanya diketahui berada di sebelah timur Papua New Guinea, dan dengan ditemukannya spesies tersebut di Yongsu menunjukkan adanya perluasan distribusi ke arah barat. Total keragaman spesies di dataran rendah ini mewakili hampir setengah dari 144 spesies yang pernah dilaporkan keberadaannya di seluruh Cagar Alam Pegunungan Cyclops, menunjukkan bahwa hutan-hutan di Yongsu memberikan kualitas habitat yang bagus untuk serangga. Kurva akumulasi spesies mengindikasikan adanya kemungkinan spesies yang ada di daerah ini tetapi tidak tercatat.

Kupu-kupu dan Ngengat: Mamberamo

Seratus dua puluh sembilan spesies kupu-kupu dari 68 genus berhasil ditemukan dengan menggunakan jaring penangkap. Jumlah tersebut mewakili sembilan famili kupu-kupu yang telah diketahui berada di Papua. Fauna kupu-kupu didominasi oleh spesies yang tipikal hidup di daerah pesisir dataran rendah Papua. Lebih dari 480 spesies ngengat mewakili lebih dari 112 genus berhasil dikumpulkan dengan menggunakan layar putih yang diberi lampu putih 160 watt. Ngengat dari famili *Pyralidae* memiliki keragaman dengan lebih dari 145 spesies, yang kebanyakan belum teridentifikasi. Ngengat besar dari famili *Sphingidae* (hanya 5 spesies) dan *Saturniidae* (tidak ada) sangat kurang terwakili. Kelimpahan dan keragaman fauna Lepidoptera di daerah Mamberamo mengindikasikan kodisi hutannya yang masih bagus. Penelitian lebih lanjut diperlukan untuk menentukan jumlah keseluruhan spesies kupu-kupu di kawasan ini, khususnya pada elevasi yang lebih tinggi, dan untuk menentukan status spesies langka dan yang dilindungi.

Ikan: Yongsu

Sebagian besar fauna ikan air tawar diperoleh dengan cara pengamatan visual menggunakan snorkel dan masker, sedang kan jaring panjang berbagai ukuran, jaring tangan, dan tuba rotenone digunakan jika perlu. Fauna ikan yang ditemukan sebanyak 33 spesies dari 25 genus dan 15 famili. Ikan-ikan di Yongsu telah teradaptasi dengan kondisi sungai yang relatif curam, dan faunanya mirip dengan kelompok yang hidup di pesisir bergunung di pesisir utara New Guinea. Fauna ikan didominasi oleh kelompok Gobi (Gobiidae dan Eleotridae) dengan jumlah hampir setengah dari total spesies yang ada. Ikan "Cling goby" sub-famili Sicydiinae terwakili dengan tujuh spesies. Daerah Yongsu dan sekitar pesisir Cyclops merupakan contoh terbaik untuk habitat sungai pesisir yang curam di seluruh Papua. Daerah ini juga merupakan rumah bagi dua ikan gobi endemik dari genus *Lentipes*, salah satunya adalah spesies baru yang ditemukan pada kegiatan pelatihan, sehingga dapat dijadikan justifikasi bagi kegiatan konservasi.

Ikan: Mamberamo

Ditemukan sebanyak 23 spesies dari 18 genus dan 11 famili dengan menggunakan jaring berbagai ukuran, pengamatan langsung, penggunaan tuba rotenone, dan mengamati ikan hasil tangkapan yang dijual di pasar Dabra. Ikan gobi kecil *Gobius tigrellus* berhasil dikoleksi untuk pertama kalinya sejak tahun 1930-an. Selama ini ikan tersebut diketahui hanya berdasarkan dari 10 spesimen yang dikumpulkan di Mamberamo oleh Ekspedisi Archbold 1938–1939. Sayangnya, enam spesies ikan introduksi juga ditemukan di sungai Mamberamo, yang memiliki proporsi spesies endemik tertinggi dibandingkan semua sungai di New Guinea. Dampak dari spesies introduksi tersebut terhadap populasi ikan asli sangat mendesak untuk segera dikaji.

Amfibia dan reptilia: Yongsu

Dua puluh enam spesies reptilia dan delapan spesies katak berhasil ditemukan di daerah Yongsu dengan mengkombinasikan pengamatan langsung dan suara serta plot serasah. Dua spesies katak tergolong spesies baru bagi ilmu pengetahuan dan satu katak lainnya terdengar dari kanopi hutan yang tinggi dan hampir dipastikan merupakan spesies *Litoria* yang belum pernah dideskripsikan. Diperlukan upaya lebih lanjut untuk mengoleksi dan mengidentifikasi spesies penghuni kanopi yang sangat menarik ini.

Lima spesies (2 katak, 3 reptilia) ditemukan di dalam plot serasah. Kurva akumulasi spesies untuk herpetofauna serasah (yang hidup di lantai hutan) mencapai garis lurus setelah 8 plot tetapi pencarian di lantai hutan dengan survei umum membuktikan bahwa herpetofauna serasah lebih banyak ditemukan daripada yang ditemukan di plot-plot tersebut. Diperlukan waktu dan usaha yang memadai untuk mencari dan membuat plot, dan teknik ini tidak menemukan spesies yang tidak terdeteksi dalam survei umum. Plot serasah tampaknya kurang efektif untuk mendapatkan inventori yang lengkap dan cepat bagi herpetofauna serasah.

Dua spesies penyu laut bertelur di pantai Yongsu: penyu hijau (*Chelonia mydas*) dan penyu tempayan (*Caretta caretta*). Tukik penyu hijau keluar dari sarangnya di camp Yongsu pada pukul 6:30 sore tanggal 25 Agustus 2000 dan seekor tukik penyu tempayan teramati di lokasi yang sama pada tanggal 28 Agustus 2000.

Amfibia dan reptilia: Mamberamo

Dua puluh satu spesies katak dan 36 spesies reptilia berhasil ditemukan di tiga lokasi dalam daerah aliran sungai Mamberamo (Furu, Tiri, dan daerah sekitar kampung Dabra) dengan cara survei langsung pada siang hari, dan survei langsung dan suara pada siang malam hari. Kura-kura/labi-labi disurvei dengan menggunakan snorkel dan masker dan juga diperoleh dari pemburu lokal. Tujuh spesies katak dan 3 spesies kadal tergolong spesies baru bagi ilmu pengetahuan. Terdokumentasinya Kadal sirip layar *Hydrosaurus amboinensis* di Furu dan Tiri menunjukkan adanya perluasan distribusi yang nyata ke arah timur untuk reptilia besar dan spektakuler ini. Beberapa reptil besar termasuk Buaya Air Tawar (*Crocodylus novaeguineae*) dan dua kura-kura air tawar—Labi-labi Raksasa (*Pelochelys cantori*) dan Kura-kura Irian (*Elseya novaeguineae*)—ditangkap oleh masyarakat lokal sebagai sumber makanan dan pendapatan. Pemanfaatan buaya di daerah ini tampaknya tidak sustainable dan penduduk kampung mengeluhkan adanya penurunan populasi pada tingkat yang mengkhawatirkan. Tingkat sustainabilitas pemanfaatan kura-kura juga perlu diprioritaskan untuk dikaji.

Burung: Yongsu

Burung didokumentasikan pada jalur bukit hutan bekas tebangan selektif dengan lereng yang sempit (40–70 m dpl) sekitar 750 m sebelah selatan camp pelatihan Yemang. Survei dilakukan selama 10 hari (20–29 Agustus 2000) dengan menggunakan jaring kabut, sensus berdasarkan suara, penghitungan pada titik tertentu, dan pengamatan secara *ad lib*. Point counts (penghitungan pada titik tertentu) mengindikasikan bahwa burung-burung hutan tersebar tidak merata pada tingkat lokal, kemungkinan akibat penyebaran sumber pakan; sensus informal pohon-pohon hutan menunjukkan bahwa tumbuhan strata arboreal tidak tersebar merata. Sembilan puluh spesies burung berhasil ditemukan, dan kebanyakan adalah burung-burung penghuni hutan. Lima belas burung tipikal hutan dataran rendah, termasuk Mambruk Victoria dan Cekakak Pita-Biasa tidak dijumpai. Sebaliknya, adanya populasi yang sehat dari Rangkong Papua, Kakatua Raja, dan Kasuari Gelambir-Satu mengindikasikan bahwa di hutan ini tidak terjadi penurunan jumlah pada fauna besarnya. Tidak adanya dataran pesisir dan hutan dataran rendah yang signifikan merupakan faktor yang memungkinkan sedikitnya jumlah spesies burung. Tidak ditemukan bukti bahwa penebangan selektif skala kecil di hutan ini membahayakan populasi burung setempat. Laporan dari naturalis setempat membuktikan bahwa burung dan mamalia di Pegunungan Cyclops masih belum banyak diteliti.

Burung: Dabra

Total sebanyak 143 spesies burung berhasil ditemukan di habitat dataran rendah, rawa-rawa dan riparian di daerah Dabra dengan menggunakan jaring kabut dan transek untuk pengamatan langsung dan suara. Dari jumlah tersebut, 65 (45%) tergolong spesies endemik New Guinea. Spesies burung yang terlihat dalam penelitian ini termasuk enam spesies burung Cenderawasih (Cenderawasih Kecil, Cenderawasih Raja, Cenderawasih Dua belas-Kawat, Manokodia Jobi, Manokodia Kilap, dan Toowa Cemerlang), Namdur (bowerbird), dan Burung Kucing (catbird). Spesies yang secara global terancam, misalnya Kasuari Gelambir Satu dan Mambruk Victoria yang sangat rentan terhadap perburuan, relatif umum dijumpai di daerah penelitian, mengindikasikan bahwa tekanan perburuan dari penduduk setempat masih rendah. Burung endemik lokal yang langka seperti Paruh-sabit Paruh Putih dan Cikukua Mamberamo tidak dijumpai selama survei, tetapi hutan rawa di sebelah utara Dabra tampaknya merupakan habitat yang cocok bagi kedua spesies tersebut sehingga perlu dilakukan survei dalam waktu dekat.

Mamalia kecil: Dabra

Mamalia kecil ditangkap dalam hutan primer sekitar Sungai Furu dan Tiri dengan menggunakan 71 perangkap Elliot (497 malam-tangkap) dan 5 jaring kabut (35 malam-tangkap). Enam puluh sembilan ekor mamalia mewakili tujuh spesies berhasil ditangkap selama survei. Mamalia yang teridentifikasi itu terdiri dari empat spesies kelelawar, dua spesies pengerat, dan satu spesies marsupial. Fauna mamalia kecil pada kedua tempat tersebut mirip; empat spesies kelelawar yang sama ditemukan pada kedua lokasi penelitian, tetapi di Sungai Tiri tidak dijumpai mamalia pengerat dan marsupial. Semua kelelawar itu tergolong kelelawar buah Megachiropteran dari famili Pteropodidae.

Satu betina *Syconycteris australis* di Furu sedang mengandung embrio, dan satu *S. australis* dan satu *Paranyctimene raptor* sedang menyusui. Delapan *Nyctimene draconilla* juga sedang mengandung embrio muda.

Ditemukannya Posum-Ekor Bulu *Distoechurus pennatus* di camp Furu berarti mengisi celah lebar dari distribusinya yang telah diketahui, dan hal ini merupakan salah satu rekor/catatan mengenai keberadaan spesies ini di Papua (Flannery, 1995). Melomis perut putih (*Melomys leucogaster*) sebelumnya hanya diketahui berada di sebelah selatan New Guinea

(Flannery, 1995) dan informasi ini merupakan catatan pertama kehadirannya di daerah utara dari Cordillera tengah. Upaya survei yang dilakukan di Dabra sangat terbatas untuk menghasilkan inveontori yang lengkap. Namun demikian, tercatatnya sebaran dua spesies tersebut diatas mengindikasikan diperlukannya survei tambahan untuk menghasilkan penemuan yang lebih menggembirakan di daerah yang belum banyak diketahui ini.

RINGKASAN KESELURUHAN DAN REKOMENDASI KONSERVASI

Komponen program pelatihan RAP 2000 di Papua pada dasarnya telah memperbaiki kapasitas lokal untuk melakukan invetarisasi keanekaragaman hayati dengan cepat. Akumulasi data yang diperoleh selama pelatihan juga telah menunjukkan pentingnya nilai biologi di hutan-hutan Yongsu dan Dabra. Khususnya penemuan beberapa spesies baru, dan perluasan daerah distribusi sejumlah spesies membuktikan betapa kurangnya informasi biologi di kawasan ini. Khusus untuk flora dan fauna di daerah aliran Sungai Mamberamo masih diperlukan sejumlah dokumentasi dan penelitian tambahan. Perlu dilakukan survei di daerah target yang ditentukan dengan seksama pada beberapa ketinggian untuk menghasilkan kajian dan penilaian yang lebih akurat mengenai pentingnya keanekaragaman hayati dan konservasi di daerah rimba utama yang luas ini. Pemahaman mengenai proses ekologi dan hubungan dalam sebuah lokasi dan antar lokasi di kawasan Mamberamo sangat diperlukan untuk memformulasikan strategi konservasi yang paling cocok untuk kawasan ini.

Daerah aliran Sungai Mamberamo menghadapi ancaman yang serius walaupun kepadatan penduduknya rendah dan hutannya masih luas. Rencana Mega Proyek Mamberamo, termasuk dam pembangkit listrik tenaga air, akan berdampak kerusakan serius bagi populasi manusia dan keragaman hayati di aliran sungai Mamberamo. Konfirmasi lebih lanjut pada daerah yang diduga memiliki keragaman hayati tinggi ini merupakan prioritas utama sehingga informasi yang akurat dapat diberikan kepada para pengambil keputusan yang terlibat dalam perencanaan proyek pembangunan di kawasan ini. Tekanan perburuan mungkin telah mempengaruhi populasi kasuari, kura-kura, dan khususnya buaya, dan tekanan ini akan bertambah seiring dengan penambahan populasi manusia dan tersedianya senjata api secara luas. Praktik perburuan yang ada sekarang ini dan spesies targetnya perlu didokumentasi sehingga dampak yang mungkin timbul dapat dikaji dan tingkat pemanfaatan yang berkelanjutan dapat dipromosikan.

Rekomendasi penelitian dan konservasi berikut ini dikembangkan berdasarkan bab-bab yang ditulis dalam laporan ini. Kebanyakan dari rekomendasi ini memerlukan kerjasama dengan penduduk setempat, LSM, dan pemerintah propinsi. Mengatasi isu introduksi spesies eksotik juga akan memerlukan proses penyadaran dan kerjasama dengan pemerintah daerah di Indonesia bagian barat.

Yongsu

- Memberikan lebih banyak kesempatan dan bantuan teknis kepada ahli biologi lokal untuk mengembangkan kemampuan ilmiahnya melalui pelatihan-pelatihan tambahan di hutan Yongsu. Pelatihan ini harus dirancang untuk memaksimalkan kerjasama dengan masyarakat setempat sehingga memberikan manfaat bagi masyarakat.

- Memantau perburuan penyu dan pengambilan telur untuk menentukan tingkat keberlanjutan dari pemanfaatannya saat ini.

- Rencana pembangunan jalan di dalam dan sekitar batas Pegunungan Cyclops harus memperhitungkan dampak yang mungkin terjadi bagi flora dan faunanya.

- Melakukan survei keragaman hayati tambahan pada lokasi yang lebih tinggi di Pegunungan Cyclops.

Mamberamo

- Meningkatkan upaya konservasi di daerah aliran Sungai Mamberamo dengan melakukan kampanye bagi komunitas nasional dan internasional untuk mempromosikan Mamberamo sebagai kawasan rimba besar yang terakhir bagi Papua.

- Melakukan penyadaran konservasi mengenai kawasan Mamberamo melalui dua pendekatan program pendidikan. Pendekatan pertama ditujukan kepada pemerintah pusat dan propinsi dan melibatkan pengembangan kurikulum yang memasukkan unsur pentingnya keanekaragaman hayati, sumber daya dan budaya di Mamberamo. Pendekatan kedua difokuskan di tingkat masyarakat dan mengembangkan program pendidikan yang dapat meningkatkan rasa bangga pada nilai biologi dan budaya Mamberamo. Program ini ditujukan bagi orang dewasa dan anak-anak sekolah di Mamberamo. Pendidikan di tingkat masyarakat juga harus mengandung unsur peningkatan keahlian masyarakat yang akan memberi nilai tambah pada kegiatan pemanfaatan sumber daya. Masyarakat lokal pada tahap awalnya memerlukan bantuan teknis untuk melakukan penilaian sumber daya, pemetaan sumber daya, dan kajian terhadap keahlian yang telah dimiliki.

- Analisa ekonomi mengenai biaya transportasi dan persyaratannya, dan biaya pengembangan pasar dan akses harus dilakukan untuk calon proyek-proyek di Mamberamo. Analisa tersebut sangat penting untuk meningkatkan proyek-proyek yang mendukung ekonomi dan ekologi dengan kerusakan sumber daya dan keragaman hayati yang minimal.

- Status populasi dan tingkat pemanfaatan buaya dan reptil air tawar besar lainnya perlu dikaji untuk memastikan keberlanjutan sumber daya ini untuk jangka panjang.

- Perlu diberikan kesempatan dan asistensi bagi ilmuwan lokal untuk melakukan penelitian biologi, sosial dan budaya di kawasan Mamberamo.

- Perlu dilakukan survei lanjutan di Mamberamo untuk mengumpulkan nilai keragaman hayati yang lebih lengkap di kawasan ini dan untuk mengkaji status spesies-spesies yang langka dan terancam. Daerah dengan prioritas tertinggi untuk RAP di masa mendatang adalah Suaka Margasatwa Foja dan Roffaer.

- Spesies eksotik (bukan asli) merupakan ancaman serius bagi hutan asli dan Perairan di Papua. Sayangnya, spesies-spesies tersebut hampir tidak mungkin untuk dimusnahkan dan kegiatan konservasi di masa mendatang harus memasukkan program pendidikan melalui penjelasan mengenai perlunya pencegahan introduksi spesies asing di masa mendatang kepada pemerintah dan lembaga lainnya yang terlibat dalam pengangkutan spesies introduksi. Efek spesies ikan introduksi terhadap populasi ikan asli perlu didokumentasikan dan informasi ini dapat digunakan sebagai bagian program pendidikan untuk mencegah terjadinya introduksi lebih lanjut di perairan ini dan perairan lainnya di Papua.

DAFTAR PUSTAKA

Conservation International. (ed.) 1999. The Irian Jaya biodiversity conservation priority-setting workshop. Final Report. Washington, DC: Conservation International.

Flannery, T.F. 1995. Mammals of New Guinea (2nd ed.). Chatswood: Reed Books.

Mack, A.L. and Alonso, L.E. (eds.) 2000. A biological assessment of the Wapoga River area of northwestern Irian Jaya, Indonesia. RAP Bulletin of Biological Assessment Number 14. Washington, DC: Conservation International.

Chapter 1

Geographic overview of the Cyclops Mountains and the Mamberamo Basin

Dan A. Polhemus and Stephen Richards

CYCLOPS MOUNTAINS

Introduction

The Cyclops Mountains are an isolated series of summits that extend along the northeastern coast of Papua from Tanah Merah Bay in the west to Jayapura in the east. The highest point, Gunung Rafeni, is 1880 m above sea level. The mountains are steep with sharp ridges and deeply incised streams, and to the north of the range many ridges extend directly to the ocean. Sandy beaches occur in a number of bays on the north coast.

The climate of the Cyclops Mountains is humid tropical. The mean annual temperature for Sentani is 26.5°C with little monthly variation. Rainfall is extremely high but there is a distinct dry season between May and October. Wettest months are December to April, when the northwest monsoon brings heavy rain to the north coast of New Guinea. Annual rainfall exceeds 5000 mm in some parts of north-coastal Cyclops Mountains (Ratcliffe, 1984).

Ratcliffe (1984) mapped the vegetation of the northern Cyclops Mountains near Yongsu as primary rainforest with small areas of secondary forest, but did not consider the narrow coastal strip examined during this course and survey. Most forest in the Cyclops Mountains grows on soils derived from metamorphic rocks, but extensive areas of basic and ultrabasic rock occur in the northeast of the mountains, and smaller areas occur in the west and possibly along the southern fringe of the range. These areas support a forest assemblage similar to that growing on metamorphic rock, but with vegetation that is distinctly stunted and typified by numerous crowded, narrow stems (Ratcliffe, 1984). A summary of the vegetation of this region is provided by van Royen (1965).

Most of the Cyclops Mountains are officially protected in the Cyclops Mountains Nature Reserve, which has a total area of 22,500 ha. However, its close proximity to Jayapura, the capital city of Papua, has exposed the reserve to intense pressure from road construction, swidden agriculture, settlements, and illegal logging. There is a demand from local communities to revise the reserve's boundaries to allow expansion of agriculture, and illegal encroachment into the reserve by local farmers and loggers is clearly visible from the southern side of the range. Maintaining the ecological integrity of the Cyclops Mountains is of paramount importance to the human population of Jayapura because the range is a significant source of fresh water for the city and surrounding settlements and forms part of the catchment for Lake Sentani. Excessive forest clearance in the reserve will increase erosion and sedimentation, reducing the quality of water supplies to Jayapura and possibly causing long-term shifts in water levels in the lake. Flash floods are more likely following removal of vegetation from the steep ridges typical of the range, and local climatic conditions, which are ameliorated by this large and heavily forested massif, are likely to shift (Ratcliffe, 1984).

Geology

The geological history of the Cyclops Mountains is complex and incompletely understood. The mountains are composed of island arc ophiolites of indeterminate age, overlain along their flanks by marine limestones (Pigram and Davies, 1987). Polhemus and Polhemus (1998) hypothesized that the Cyclops Mountains might represent one of the most westerly of the accreted terranes derived from the Solomons Arc system as defined by Kroenke (1984), but noted that this was speculative in the absence of detailed age and petrology data. More recent evidence (Davies, pers. comm) indicates that the Cyclops Mountains may be a terrane derived from the earlier Melanesian Arc, which accreted to central New Guinea from the late Cretaceous through the early Oligocene (Kroenke, 1984). If true, then the anomalous position of the Cyclops Mountains terrane on the northern coast of central New Guinea, rather than deeper inland along the northern flanks of the central mountains as is typical of other Melanesian Arc terranes, may indicate that this terrane collided with the apex of a salient of underlying continental basement that projects northward along the Papua-Papua New Guinea border region (Davies, 1990; Polhemus and Polhemus, 1998).

In either case the Cyclops Mountains represent a former island that was incorporated into the greater land mass of New Guinea some time in the Tertiary. Initial colonization, diversification, and evolution of its flora and fauna would therefore have occurred in isolation from the main body of New Guinea, with the eventual collision producing a composite biota containing a mixture of local Cyclops endemics intermingled with mainland New Guinea species that invaded the terrane following accretion.

Previous expeditions

Though the first biological specimens were collected from the Cyclops Mountains as early as 1889, there have been few attempts to systematically survey the area's flora and fauna. A major study on the region's birds was undertaken by Ernst Mayr in 1928, and in 1936 the intrepid naturalist-explorer J. Evelyn Cheesman made extensive collections of insects and other taxa, reaching elevations in excess of 1050 m (Cheesman, 1938). Members of the Third Archbold Expedition to New Guinea collected along the foothills of the Cyclops Mountains in June and July of 1938 while awaiting the establishment of field camps further south in the Mamberamo Basin. Limited collections and observations were made at elevations up to 900 m. a.s.l. during short forays from Jayapura (then Hollandia) by the expedition's specialists in entomology (L.J. Toxopeus and J. Olthof), botany (L. J. Brass), mammology (W. B. Richardson), ornithology (A. L. Rand), and forestry (E. Meyer-Drees). A short account of these activities is presented in Archbold et al. (1942). Despite these various survey efforts, and the proximity of the range to a research university in modern Jayapura, the biota of the Cyclops Mountains remains under-documented; indeed, the recent discovery of a new but possibly extinct species of Echidna in the Cyclops (Flannery and Groves, 1998) indicates that the range continues to harbour many undiscovered biological treasures.

Description of study site

Training and survey at Yongsu were conducted in a narrow strip of lowland and hill forest between the mountains and the ocean. The topography is steep, rising rapidly to altitudes of >1000 m asl within about 5 km of Yemang Camp. A large stream falling steeply from the north slopes of the mountains reaches the sea adjacent to Yongsu Dosoyo, and several other smaller streams reach the ocean at Yemang and Jari. Activities were concentrated around Yemang camp and in Jari forest. Most of the area has a cover of primary rainforest but secondary forest occurs in patches around the Yemang Camp, and the flat valley between Yongsu Dosoyo and the base of the mountains is intensively gardened.

MAMBERAMO BASIN

Introduction

The Mamberamo River is the largest in northern Papua, draining a catchment that encompasses all northward flowing streams descending from the New Guinea central mountains between the Papua New Guinea border and approximately 137° W longitude. The catchment is shaped like a giant inverted "T," with two major branches, the Rouffaer in the west and the Idenburg in the east, flowing along roughly east-west courses, then meeting in the Meervlakte Basin to form the main Mamberamo River.

The upper Rouffaer branch rises at elevations above 4000 m in the Nassau Range, with most of this headwater drainage feeding into rivers that follow the east-west strike of the Derewo Fault Zone. The two most significant of these drainages are the Delo, or Hitalipa, in the west, and the upper Rouffaer itself in the east; these two branches meet at a confluence below the rugged peak of Mt. Gulumbulu (4041 m), which marks the ancient boundary between the country of the Moni people to the west and the Dani people to the east. Below this confluence the Rouffaer flows north through a gorge for approximately 50 km, then turns east as it enters the Meervlakte Basin, where it receives several large tributaries from its southern bank, including the Van Daalen and Swart Rivers in its final 125 km before joining the Idenburg.

The Idenburg River rises in poorly mapped country north of Puncak Mandala (4700 m), one of the highest peaks in Papua. It is an area of fractured karst topography with many structurally controlled drainages such as the Kloof, Borme and Sobger rivers which form incised, reticulate networks with the main trend of flow toward the northwest. Near the village of Huluatas this complex of rivers coalesces with southward flowing drainages from the flanks of the 30 and 60 Mile Hills between Lake Sentani and the Meervlakte Basin, forming the westward flowing Idenburg River. Beyond Huluatas the Idenburg flows across the Meervlakte Basin in sinuous bends for nearly 200 km, receiving several major south bank tributaries. Most notable is the Van de Wal, which drains the eastern limb of the Derewo Fault Zone and contains Archbold Lake in its upper reaches.

Beyond the confluence of the Rouffaer and the Idenburg, the Mamberamo turns abruptly northward, flowing for 175 km to the coast at Cape D'Urville on the northeast margin of Cenderawasih Bay. The lower reaches of the river pass through a deep gorge in the Foja/Van Rees mountain chain containing several sets of navigable rapids, the most famous of these being the Batavia Rapids, and also skirts the large and poorly investigated Lake Rombebai on the northern coastal plain.

Human population density along the entire length of the Mamberamo and its combined tributaries is exceptionally low, and as a result the Mamberamo Basin is one of the largest remaining tropical wilderness areas on earth. The absence of dams in its mountainous headwater reaches and lack of significant agricultural development along its lowland floodplains marks the Mamberamo as one of the few nearly pristine tropical rivers remaining in the world. In addition, the associated Meervlakte Basin and Mamberamo Delta represent some of the largest existing undisturbed tropical wetland systems. The value of this system in terms of biological conservation is obvious, and if proposals to dam the river at its point of passage through the Foja/Van Rees Mountains proceed to fruition it will represent an ecological catastrophe for Papua's biodiversity.

Current Legal Status and Conservation Efforts

Within the Mamberamo Basin, the Indonesian government declared and established the Sungai Rouffaer Wildlife Sanctuary (310,000 ha) in 1980 and the Pegunungan Foja Wildlife Sanctuary (1,018,000 hectares) in 1982. Located in the middle of the Mamberamo Basin, the elevation of these sanctuaries ranges from sea level up to 2,100 meters. Both sanctuaries were designated to protect representative ecosystems of Papua's distinct northern lowlands, the Foja mountain flora and fauna, and the nesting sites of the threatened New Guinean crocodile, *Crocodylus novaeguinea*. They also provide protection to over 150 bird species, including five species of birds of paradise and Brass's Friarbird, *Philemon brassi*.

Although both reserves have been officially designated and gazetted, no management activities exist on the ground. Neither sanctuary has a management plan or sufficient management infrastructure. The National Conservation Plan for Indonesia (Ministry of Forestry, 1995) recommended that both areas should be merged and extended and then declared as a National Park, for a total area of 1,749,200 ha.

The majority of Mamberamo communities are still involved in subsistence hunting, fishing in the Mamberamo River, and small-scale agriculture. Some communities are exploring other income sources, such as commercial crocodile hunting and collection of gaharu resin. In general, local communities still depend strongly on their land and the surrounding natural resource base.

Through working with local partners, Conservation International aims to demonstrate a holistic approach to maintaining intact biodiversity in the Mamberamo Basin. This approach seeks to balance economic development with the sustainable use of natural resources so valued by local communities. CI is working in collaboration with a local community group, the Mamberamo Adat Council, who desire to protect their land from extractive industries and to develop a community reserve that would include several zones designated for daily use, hunting, culture, and protection of core areas and biodiversity.

As an alternative to large-scale industrial development plans, Conservation International is assisting local and provincial partners to design an alternative development plan for the great Mamberamo watershed—one that takes into account the development needs of the local Mamberamo populace and the ecological sustainability of this fragile ecosystem. CI's current vision includes a Basin-wide land-management exercise, to be conducted by government, nongovernment, and indigenous peoples' representatives. This will build on the considerable social and economic assessments that have been carried out by CI's Resource Economics Department in the form of the Papua *Resource and Conservation Economics* analysis. The ultimate goal is a resource-use plan that will meet the economic needs of the local residents of the Basin, provide an opportunity to conserve substantial forest and biodiversity of the Basin, and permit appropriate province-wide development in a way that does not threaten the Basin's critical natural values. From a conservation viewpoint, this overall activity will result in the creation of the Mamberamo Raya Corridor, linking the major natural features of the basin into a network of protected landscapes that will conserve watersheds, local livelihoods, and biodiversity. To complete such a complex program of field conservation, CI expects to carry out a series of additional studies, including additional biological surveys and environmental awareness-building. This initial Mamberamo RAP has fostered considerable interest and pride within the local stakeholders, who now better understand how remarkable their forest home is.

Previous expeditions to the Dabra region

The first Europeans to reach the Dabra area were part of the Opperman Expedition, a Dutch military survey whose primary aim was a reconnaissance of the Van Rees Mountains and coastal islands near the mouth of the Mamberamo. In November 1914, a detachment from this expedition, commanded by L. A. C. M. Doorman, made a foray upstream along the Mamberamo in an attempt to reach the central mountains. Turning east along the Idenburg, they followed this river to the point where it makes its first close approach to the outlying foothills. Here they ascended a swift, clear river entering from the south bank, eventually reaching its headwaters on a mountaintop at 3580 m. The river was later named Doorman River, and the mountain at its head Doorman Top.

In 1920 another large Dutch expedition, with 800 men under the command of Captain J. H. G. Kremer, returned to this same area with the intention of penetrating further into the central ranges. This party came by boat up the now-familiar route along the Mamberamo and Idenburg rivers, establishing a camp called Prauwen-bivak just upstream from the mouth of the Doorman River on the Idenburg, near present day Dabra. Certain parties of this expedition ascended once again to Doorman Top, then continued southward into the deep valley of the Swart River, which drains into the Rouffaer, the major western branch of the Mamberamo system. From here they pushed southward again, over high and rugged terrain into the headwaters of the Baliem River. Although advance parties would eventually summit Puncak Trikora (at that time known as Wilhelmina Top), the expedition entirely missed the Baliem Valley itself, which lay to the east of their route. Despite spending nearly a year and a half in the Dabra area and adjacent mountains, Kremer's large expedition generated remarkably few noteworthy scientific results, outside of an excellent botanical account, including a map of the region, by Lam (1945). As a result, the current RAP survey was still the first expedition to make any systematic biological collections from this area.

The 3rd Archbold Expedition (Archbold et al., 1942) made an aerial reconnaissance of the Doorman River and its environs in late June 1938, using their PBY float-plane named the "Guba" to determine whether the river offered a feasible route into the higher mountains. They characterized the area as "...rough, forbidding country, a succession of deep, dark valleys and forested ridges, each several thousand feet high...". This expedition instead chose to locate their base camp further eastward on the Idenburg, at a site they called Bernhard Camp, about 60 kilometers upstream from the Prauwen-bivak of the Kremer Expedition. Local people at Dabra still claim to know the location of this camp, which they indicate can be reached by motorized canoe in 8 hours. Bernhard Camp was on a bluff overlooking a backwater of the river at 50 m elevation, and was some 9 m above the water when established in late June 1938. By March 1939, at the height of the rainy season (Table 1.1) this camp had been completely flooded to a depth of 0.3 m even on the highest ground, demonstrating the vast fluctuations in seasonal rain-

fall and corresponding river levels prevailing in this section the Mamberamo Basin.

From this base the expedition worked into the mountains fronting the Meervlakte, establishing a series of increasingly higher scientific collecting camps (Table 1.2) that culminated with Top Camp at 2150 m. Rather than continue overland into the valley of the Hablifuri River through progressively more rugged country, this expedition instead used its float plane to shuttle expedition members from Bernhard Camp to a camp on Lake Habbema, near the base of the Puncak Trikora massif, from which investigations on the higher elevation plant communities could be easily conducted. In the course of these transits the expedition discovered the remarkable upland Dani communities of the Baliem Valley, as well as another lake at middle elevation in the upper Hablifuri catchment. This second lake, named Lake Archbold, was used as a landing site for the float plane but was not used as a base for scientific collecting activities. In terms of its scope, organization, and subsequent scientific publications, the

Table 1.1. Rainfall data for Bernhard Camp, Mamberamo River Basin, Papua, 1938–1939 (Archbold et al., 1942).

Month	Rainfall (mm)
July 1938 (21 days)	100
August 1938	259
September 1938	718
October 1938	304
November 1938	547
December 1938	454
January 1939	561
February 1939	511
March 1939	539
April 1939	802
Total (10 months)	**4795**

Table 1.2. 1938–1939 Archbold Expedition camp locations (Archbold et al., 1942).

Camp Name	Elevation (m)	Coordinates (Approx)
Bernhard Camp	50	3°29'S, 139°13'E
Araucaria Camp	850	3°30'S, 139°11'E
Rotan Camp (Tusschencamp)	1200	3°30'S, 139°09'E
Mist Camp	1800	3°30'S, 139°05'E
Top Camp	2150	3°30'S, 139°02'E
Bele River Camp (Ibèlè Camp)	2200	N/A
Mosbosch Camp	2800	N/A
Brievenbus Camp (Letterbox Camp)	3560	4°13'S, 138°44'E
Puindal Camp	3800	N/A

3rd Archbold Expedition was the most significant scientific exploration ever conducted in the Mamberamo Basin, and its results, particularly in terms of the floral and faunal collections from Bernhard Camp and the adjacent foothills, stand as an impressive benchmark against which those from the current RAP survey may be compared.

Following the 3rd Archbold Expedition and the subsequent outbreak of World War II, little additional scientific work was done in the Dabra area. In 1995 the P. T. Freeport Indonesia mining company conducted mineral exploration activities in the Doorman River catchment and surrounding regions, using Dabra as a staging point for supplies that were lifted to drill camps in the mountains by helicopter. No biological collections were made during these surveys, and no subsequent mining activity was pursued in the region.

In terms of scientific history, the Dabra area has clearly been one of the more active regions in north central New Guinea, with many expeditions passing through the area on their way to the mountains south of the Idenburg. It is unfortunate that this succession of expeditions did not produce a commensurate accumulation of biological specimens, and with the exception of the 3rd Archbold Expedition's collections from Bernhard Camp and environs to the east, little was known of the overall natural history of the Dabra area prior to the current RAP expedition.

In regard to the original inhabitants of the Mamberamo Basin at the time of these first outside contacts, the Archbold report (Archbold et al., 1942) notes "The Meervlakte hereabouts was inhabited by a scattered nomadic people...they called themselves the Tabbertoea. In times of high flood they took to their canoes with their dogs and a few household belongings and disappeared from the neighborhood, to return when the waters had somewhat receded. Van Arcken's description of the men as 'big, heavily built, wild-looking fellows, covered thickly with itch' (i.e., ringworm) fitted well. Apparently they planted nothing, unless they can be credited with establishing some of the numerous breadfruit trees which occurred on the banks of the waterways. They built only temporary houses, subsisted on sago and by hunting and fishing, and perhaps seldom ventured far from their clumsy dugouts."

The native canoes also attracted the attention of Lam (1945), who passed through the region in 1920 and commented on "...the peculiar proas such as we saw only on the Meervlakte. They are hewn out of a heavy trunk, the bottoms broad, the body rather deep, the sides sloping inward and supplied with narrow openings, the upper part narrow. The bow and stern are cut off transversely, therefore, seen from the front, they appear to be blunt; seen from above, however, they seem to be narrow. Perhaps this form was developed because of the unfriendly attitude of different tribes along the river. The crew can conceal itself entirely behind the sides and keep an eye on the opponents through the holes."

Such canoes now seem to be a thing of the past on the Mamberamo and its tributaries, since all of those seen during the current survey were of the typical low-sided dugout variety common throughout modern New Guinea. In addition, most of the population of the region, which still amounts to very few people, is now concentrated around a few permanent towns with airstrips, such as Dabra. The recent nature of this change can be judged by the fact that the settlement of Dabra did not exist when the 3rd Archbold Expedition passed through the area between 1938 and 1939.

Description of the Dabra RAP sites

The Dabra area represents an abrupt topographic and geological transition from the flat, swampy plains of the interior Mamberamo Basin to the steep northern slopes of the New Guinea central mountains. The Mamberamo lowlands consist of deep deposits of Quaternary alluvium eroded from the adjacent mountain flanks, which at Dabra represent the northern margin of an accreted Eocene island arc complex that exhibits, from north to south, sequential belts of mid to late Tertiary volcanics, Eocene ophiolite, and Eocene metamorphics (Dow et al., 1986). The study sites for the current RAP lay primarily within the volcanic belt, the rocks of which have not been accurately dated but are probably Eocene given their association with the remainder of the arc complex. There were evident similarities between the Dabra volcanics and those seen at the Logari River site surveyed on the Wapoga RAP in 1998 (Mack and Alonso, 2000), in particular the presence of a distinctive black hornfels in the streambed cobble bars. The Furu River site had a less diverse mixture of rock types than the Tiri River; at the former site the bed substrates were composed entirely of volcanic materials, while at the latter site the bed materials included examples of ophiolitic and metamorphic rocks, including marble, that indicated its headwaters penetrated far back into the mountains, evidently to the margin of the metamorphic belt.

The geological transition present at Dabra is also reflected ecologically in the distribution of forest types and their associated aquatic habitats. The Quaternary alluvium of the Mamberamo lowlands is covered by extensive swamp forests with networks of lakes and still water channels, similar in many respects to those sampled on the lower Wapoga River below Siewa in 1998. Whereas at Siewa there was an extensive zone of terre firme lowland forest lying between these swamp forests and the edge of the mountain foothills, at Dabra the transition from swamps to foothills was abrupt, with the terre firme lowland forests occupying only a narrow zone at the extreme base of the hills, or in slightly elevated embayments such as the Doorman River catchment. The Tiri River, lying in the latter area, had braided streambeds on sand and gravel outwash similar to those seen in the Siewa area during the 1998 RAP, while the Furu River, by contrast, made a quick transition to a rocky, high gradient bed similar to streams in the Logari area. This foreshortening of aquatic habitat zones along the Furu reflects the fact that in this latter area the volcanic belt of the outer foothills slopes steeply into the swampy alluvial plains, with a limited pediment of outwash sediments on which the lowland forests and streams typical of them can develop. A more extensive discussion of these patterns of forest zonation along the Mamberamo can be found in Archbold et al. (1942).

LITERATURE CITED

Archbold, R., A.L. Rand, and L.J. Brass. 1942. Results of the Archbold Expeditions. No. 41. Summary of the 1938–1939 New Guinea Expedition. Bull. Am. Mus. Nat. Hist. 79: 197–288.

Cheesman, L.E. 1938. The Cyclops Mountains of Dutch New Guinea. Geog. J. Lond. 91: 21–30.

Davies, H. L. 1990. Structure and evolution of the border region of Papua New Guinea. *In:* Petroleum Exploration in Papua New Guinea: Proceedings of the First PNG Petroleum Convention, Port Moresby, 12–14th February 1990. G. J. Carman and Z. Carman (eds.). Papua New Guinea: PNG Chamber of Mines and Petroleum.

Dow, D.B., G.P. Robinson, U. Hartono, and N. Ratman. 1986. Geologic map of Irian Jaya, Indonesia. Geological Research and Development Centre, Ministry of Mines and Energy, Bandung. 1:1,000,000 scale map.

Flannery, T.F. and C.P. Groves. 1998. A revision of the genus *Zaglossus* (Monotremata, Tachyglossidae), with description of new species and subspecies. Mammalia 62: 367–396.

Kroenke, L.W. 1984. Cenozoic development of the Southwest Pacific. United Nations Economic and Social Commission for Asia and the Pacific, Committee for Co-ordination of Joint Prospecting for Mineral Resources in South Pacific Offshore Areas, Technical Bulletin 6.

Lam, H.J. 1945. Fragmenta Papuana [Observations of a naturalist in Netherlands New Guinea]. Sargentia, 5: 1–196.

Ministry of Forestry. 1995. National Conservation Plan for Indonesia: Vol. 9. Irian Jaya Biogeographic Region and Province. Ministry of Forestry, Jakarta.

Pigram, C.J. and H.L. Davies. 1987. Terranes and the accretion history of the New Guinea orogen. Bureau of Mineral Resources, J. Aust. Geol. Geophys. 10: 193–211.

Polhemus, D. and J.T. Polhemus. 1998. Assembling New Guinea: 40 million years of island arc accretion as indicated by the distributions of aquatic Heteroptera (Insecta). *In:* Hall, R. and J.D. Holloway (eds.). Biogeography and geological evolution of SE Asia. Leiden: Backhuys. Pp. 327–340.

Ratcliffe, J.B. 1984. Cagar Alam Pegunungan Cyclops Irian Jaya: Management Plan 1985–1989. Jayapura: World Wildlife Fund.

van Royen, P. 1965. Sertulum Papuanum 14. An outline of the flora and vegetation of the Cyclops Mountains. Nova Guinea N.S. Botany. 21: 451–469.

Bab I

Ulasan Geografis Pegunungan Cyclops dan Daerah Aliran Sungai Mamberamo

Dan A. Polhemus dan Stephen Richards

Pengantar

Pegunungan Cyclops merupakan serangkaian puncak-puncak gunung yang terisolasi, membentang di sepanjang pantai timur laut Papua, dari Teluk Tanah Merah di sebelah barat ke Jayapura di sebelah timur. Puncak paling tinggi—Gunung Rafeni—adalah 1880m dpl (dari permukaan laut). Pegunungan Cyclops bertopografi curam dengan celah-celah yang tajam dan aliran-aliran air yang terpilah-pilah dengan jelas. Ke arah utara dari barisan pegunungan tersebut terdapat celah-celah yang membentang langsung ke lautan. Dapat ditemukan pantai-pantai berpasir pada sejumlah teluk di pesisir utara.

Pegunungan Cyclops memiliki iklim tropis lembab. Rata-rata suhu tahunan untuk Sentani adalah 26,5º Celcius dengan variasi bulanan yang kecil. Curah hujan sangat tinggi, tetapi terdapat musim kering yang jelas antara bulan Mei dan Oktober. Bulan paling basah adalah Desember sampai April, ketika angin Barat Laut membawa banyak hujan ke pantai utara New Guinea. Curah hujan tahunan melebihi 5000 mm di beberapa bagian pantai utara Pegunungan Cyclops (Ratcliffe, 1984).

Ratcliffe (1984) memetakan vegetasi bagian utara Pegunungan Cyclops sekitar Yongsu sebagai hutan hujan primer dengan sejumlah kecil areal hutan sekunder, tetapi tidak memasukkan jajaran pantai sempit yang diteliti selama kursus dan survei ini. Kebanyakan hutan di Pegunungan Cyclops, tumbuh di tanah yang berasal dari bebatuan metamorfik, tetapi sejumlah besar areal bebatuan dasar dan ultrabasic terdapat di timur laut Pegunungan Cyclops, dan sejumlah kecil areal di sebelah barat dan kemungkinan di sepanjang celah-celah bagian selatan. Kawasan tersebut mendukung gugusan hutan yang mirip dengan hutan yang tumbuh di bebatuan metamorfik, tetapi dengan vegetasi yang berbeda dan dilingkupi oleh tanaman merambat yang padat (Ratcliffe, 1984). Rangkuman mengenai vegetasi wilayah ini diberikan oleh van Royen (1965).

Sebagian besar kawasan Pegunungan Cyclops berstatus dilindungi sebagai Cagar Alam Pegunungan Cyclops, dengan luas total 22,500 Ha. Namun demikian, lokasinya yang dekat dengan Jayapura, ibu kota Papua, mengakibatkan cagar alam ini mendapat tekanan kuat dari adanya pembangunan jalan, kebun-kebun, pemukiman, dan penebangan hutan ilegal. Ada permintaan dari masyarakat lokal untuk merevisi batas-batas cagar alam untuk perluasan wilayah pertanian. Perambahan ilegal ke dalam wilayah cagar alam oleh petani-petani lokal dan para penebang hutan terlihat jelas dari sisi sebelah selatan pegunungan ini. Mempertahankan integritas ekologis Pegunungan Cyclops adalah kepentingan paling pokok bagi penduduk di Jayapura karena wilayah ini merupakan sumber air tawar yang penting bagi kota dan pemukiman-pemukiman di sekitarnya dan membentuk bagian dari resapan air untuk Danau Sentani. Pembukaan hutan berlebihan di cagar alam akan meningkatkan erosi dan sedimentasi, mengurangi kualitas air yang disalurkan ke Jayapura dan kemungkinan dalam jangka panjang menyebabkan perubahan tinggi permukaan air danau. Kemungkinan terjadinya banjir bandang akan lebih besar akibat menghilangnya vegetasi dari celah-celah yang curam di wilayah pegunungan ini, dan kondisi iklim lokal yang bagus karena adanya hutan lebat, tampaknya akan berubah (Ratcliffe, 1984).

Geologi

Sejarah geologi Pegunungan Cyclops tergolong kompleks dan belum sepenuhnya dipahami. Pegunungan ini terdiri atas Ophiolite yang umurnya tidak diketahui, dan batu kapur lautan di sepanjang celahnya (Pigram dan Davis, 1987). Polhemus dan Polhemus (1998), memberikan hipotesa bahwa Pegunungan Cyclops kemungkinan mewakili satu bagian paling barat dari dataran tambahan yang berasal dari sistem Busur Solomon sebagaimana yang didefenisikan oleh Kroenke (1984), tapi perlu dicatat bahwa ini merupakan spekulasi karena tidak adanya data lebih detil mengenai umur dan unsur minyak (petrology). Bukti terbaru (Davies, Komunikasi pribadi) mengindikasikan bahwa Pegunungan Cyclops kemungkinan adalah sebuah dataran yang berasal dari Busur Melanesia awal yang kemudian ditambahkan ke bagian tengah New Guinea sejak akhir jaman Cretaseus sampai awal Oligocene (Kroenke, 1984). Jika benar, maka posisi yang aneh dari dataran Pegunungan Cyclops di pantai utara New Guinea tengah, dan bukannya dataran yang lebih dalam di sepanjang celah bagian utara dari pegunungan tengah, merupakan ciri dari dataran-dataran busur Melanesia, mungkin mengindikasikan bahwa lempengan ini bertabrakan dengan puncak dari dasar benua yang mendorong ke arah utara di sepanjang perbatasan Papua-Papua New Guinea (Davies, 1990; Polhemus dan Polhemus, 1998).

Terlepas dari kedua kemungkinan tadi, Pegunungan Cyclops mewakili apa yang sebelumnya merupakan sebuah pulau yang bergabung dengan dataran besar New Guinea pada masa tertier. Kolonisasi awal, diversivikasi, dan evolusi flora fauna kemungkinan terjadi dalam situasi terisolasi dari bagian utama New Guinea. Selanjutnya dengan adanya tabrakan menghasilkan kumpulan biota yang terdiri atas campuran species endemik lokal Cyclops dengan species-species daratan utama New Guinea yang telah ada sebelumnya.

Ekspedisi-ekspedisi terdahulu

Walaupun spesimen-spesimen biologi awal sudah dikumpulkan dari Pegunungan Cyclops sejak tahun 1889, baru sedikit usaha yang secara sistematis mensurvai flora dan fauna wilayah ini. Sebuah studi berskala besar tentang burung-burung di wilayah ini dilakukan oleh Ernst Mayr pada tahun 1928, dan pada tahun 1936 J. Evelyn Cheesman (seorang penjelajah alam yang pemberani) melakukan koleksi yang menyeluruh terhadap serangga-serangga dan taksa lain, pada ketinggian lebih dari 1050 m (Cheesman 1938). Anggota dari Ekspedisi Archbold III ke New Guinea mengoleksi berbagai spesimen di sepanjang kaki bukit Pegunungan Cyclops pada bulan Juni dan Juli 1938 sementara menunggu pembangunan camp (kemah) di selatan Daerah Aliran Sungai Mamberamo. Sejumlah pengamatan dan koleksi terbatas dilakukan pada ketinggian 900 m dpl dalam sebuah studi singkat dari Jayapura (dulu bernama Hollandia) oleh para ahli serangga (L.J. Toxopeus dan J. Olthof), botani (L.J.

Brass), mamalia (W.B. Richardson), burung (A.L. Rand) dan kehutanan (E. Meyer-Drees). Aktivitas aktivitas ini dipresentasikan oleh Archbold dkk. (1942). Terlepas dari berbagai usaha survai tersebut, dan dekatnya lokasi dengan Universitas Cenderawasih, biota Pegunungan Cyclops masih kurang terdokumentasi; bahkan penemuan terakhir berupa species baru Echidna yang mungkin telah punah di Cyclops (Flannery dan Groves, 1998) mengindikasikan bahwa wilayah ini masih menyimpan banyak kekayaan biologi yang belum ditemukan.

Gambaran Lokasi Studi

Pelatihan dan survei di Yongsu dilakukan di sebuah dataran rendah yang sempit dan hutan perbukitan di antara pegunungan dan laut. Topografinya curam, naik dengan tajam ke ketinggian > 1000 m dpl dalam jarak 5 km dari Camp Yemang. Aliran sungai yang curam dari dari lereng utara Pegunungan menuju ke laut sekitar Yongsu Dosoyo, dan beberapa aliran sungai kecil juga bermuara ke laut sekitar Yemang dan Jari. Aktivitas – aktivitas dikonsentrasikan di sekitar camp Yemang dan hutan Jari. Kebanyakan areal tersebut memiliki tutupan hutan hujan primer tetapi hutan sekunder terdapat pada beberapa bagian di sekitar camp Yemang, sedangkan lembah datar antara Yongsu Dosoyo dan kaki Pegunungan Cyclops telah menjadi kebun.

DAERAH ALIRAN SUNGAI MAMBERAMO

Pengantar

Sungai Mamberamo adalah yang paling besar di utara Papua, menampung seluruh aliran sungai kecil yang mengalir ke utara dari pegunungan tengah New Guinea antara perbatasan Papua New Guinea dan kira-kira 137° derajat Bujur Barat. Aliran sungai tersebut berbentuk seperti huruf "T" besar terbalik, dengan dua cabang utama, Rouffaer di bagian barat dan Idenburg di bagian timur, mengalir ke arah timur-barat, lalu bertemu dengan lembah Sungai Meervlakte yang kemudian menjadi Sungai Memberamo.

Bagian hulu dari anak Sungai Rouffaer berada pada ketinggian di atas 4000 m di wilayah Nassau dan sebagian besar dari hulunya masuk ke sungai-sungai yang mengikuti timur-barat dari Zona Derewo Fault. Dua aliran air yang paling penting adalah Delo, atau Hitalipa di sebelah barat dan hulu Rouffaer di sebelah timur. Kedua anak sungai ini bertemu di bawah puncak Gunung Gulumbulu 4041 m, yang merupakan tanda batas pada masa lalu antara Suku Moni di barat dan Suku Dani di timur. Di bawah pertemuan dua sungai tadi, Sungai Rouffaer mengalir ke utara kira-kira sepanjang 50 km melalui sebuah ngarai, kemudian berbelok ke timur sewaktu memasuki lembah Sungai Meervlakte. Di lembah ini Sungai Rouffaer menerima aliran air dari sejumlah anak sungai - anak sungai yang besar dari arah selatan, termasuk dari Sungai Van Daalen dan Sungai Swart di 125 km terakhir, sebelum bergabung dengan Sungai Idenburg.

Sungai Idenburg berada di daerah yang tidak terpetakan dengan baik di sebelah utara Puncak Mandala (4700 m), salah satu puncak tertinggi di Papua. Wilayah ini memiliki topografi dengan pecahan batuan karst dan memiliki sumber-sumber air yang dikontrol secara struktural seperti Sungai Kloof, Borme dan Sobger yang membentuk aliran yang berpilah-pilah dan jaringan ber-retikulasi dengan kecenderungan aliran ke arah barat laut. Di dekat Desa Huluatas, kelompok sungai-sungai ini bergabung dengan aliran air yang mengalir ke selatan yang berasal dari celah-celah pegunungan Bukit Mile 30 dan 60 antara danau Sentani dan lembah Sungai Meervlakte. Pertemuan tersebut membentuk aliran ke arah barat Sungai Idenburg. Di luar Desa Huluatas, Sungai Idenburg mengalir melalui lembah Sungai Meervlakte dengan lengkungan berkelok-kelok sejauh 200 km dan menerima limpahan air dari beberapa anak sungai di tepi sungai bagian selatan. Terutama Sungai Van de Wal yang mengalir ke cabang bagian timur dari Zona Derewo Fault dan memiliki Danau Archbold di bagian atasnya.

Di luar pertemuan Sungai Rouffaer dan Sungai Idenburg, Sungai Mamberamo berbalik dengan tajam ke arah utara, dan mengalir sejauh 175 km ke arah pantai di Tanjung D'Urville yang terletak di sisi timur laut Teluk Cenderawasih. Bagian bawah dari sungai melewati ngarai yang dalam di rangkaian pegunungan Foja/Van Rees yang memiliki beberapa arus deras yang bisa dilalui, yang paling terkenal adalah arus deras Batavia, dan juga danau Rombebai yang besar dan belum banyak diteliti, terletak di dataran pesisir utara.

Kepadatan penduduk di sepanjang Sungai Mamberamo dan anak-anak sungainya sangat rendah, sehingga Daerah aliran Sungai Mamberamo merupakan salah satu daerah rimba utama terluas di bumi. Tidak adanya dam/waduk di hulu sungai dan pengembangan pertanian yang signifikan di dataran rendahnya membuat Mamberamo sebagai satu dari sedikit sungai daerah tropis yang masih hampir murni yang masih ada di dunia. Selain itu, penggabungan lembah Sungai Meervlakte dan Delta Mamberamo mewakili beberapa sistem lahan basah tropis terbesar yang tidak terganggu. Nilai dari sistem ini dalam hal konservasi biologi adalah jelas, dan jika usulan-usulan untuk membendung sungai di daerah di mana alirannya melalui Pegunungan Foja/Van Rees dilanjutkan untuk dilaksanakan, maka hal itu akan merupakan contoh sebuah bencana ekologi bagi keanekaragaman hayati Papua.

Status Hukum dan Upaya Konservasi Saat ini

Di daerah Aliran Sungai Mamberamo, Pemerintah Indonesia telah menetapkan Suaka Margasatwa Sungai Rouffaer (310,000 ha) pada tahun 1980 and Suaka Margasatwa Pegunungan Foja (1,018,000 hectares) pada tahun 1982. Terletak di tengah-tengah kawasan Mamberamo dengan ketinggian mulai dari permukaan laut hingga 2.100 meter. Penentuan kedua Suaka Margasatwa tersebut bertujuan melindungi perwakilan ekosistem dataran rendah sebelah utara Papua yang berbeda, flora dan fauna Pegunungan Foja, dan tempat bersarang Buaya New Guinea yang terancam, Crocodylus

novaeguinea. Kawasan itu juga untuk melindungi lebih dari 150 spesies burung, termasuk lima spesies burung Cenderawasih dan Cikukua Mamberamo, Philemon brassi.

Walaupun kedua suaka tersebut secara resmi telah ditentukan dan diumumkan, belum ada aktivitas pengelolaan di lapangan. Keduanya belum mempunyai rencana pengelolaan atau infrastruktur pengelolaan yang memadai. Dalam National Conservation Plan for Indonesia (Departemen Kehutanan, 1995), kedua daerah itu direkomendasi untuk digabung dan diperluas, kemudian ditetapkan sebagai Taman Nasional dengan luas total 1,749,200 hectares.

Sebagian besar masyarakat Mamberamo masih melakukan perburuan subsisten, mencari ikan di Sungai Mamberamo, dan pertanian skala kecil. Ada juga masyarakat yang memperoleh pendapatan dari berburu buaya dan mengumpulkan gaharu. Secara umum, masyarakat lokal masih sangat bergantung pada lahan mereka dan sumberdaya di sekitarnya.

Bekerjasama dengan mitra lokal, Conservation International ingin mendemonstrasikan pendekatan yang menyeluruh (multi dimensi) untuk memelihara keutuhan keragaman hayati di daerah aliran sungai Mamberamo. Pendekatan ini berupaya menyeimbangkan pengembangan ekonomi dengan pemanfaatan yang berkelanjutan oleh masyarakat. CI saat ini bekerjasama dengan kelompok masyarakat lokal, Dewan Masyarakat Adat Mamberamo, yang ingin melindungi tanah mereka dari industri-industri ekstraktif dan untuk mengembangkan hutan lindung masyarakat yang akan mencakup beberapa zona untuk pemanfaatan, perburuan, budaya, serta perlindungan terhadap daerah inti dan keragaman hayatinya.

Sebagai alternatif rencana pembangunan industri skala besar (dan tidak praktis), Conservation International sedang membantu para pihak lokal untuk merancang rencana pembangunan alternatif di daerah aliran sungai Mamberamo--satu hal yang harus diperhatikan adalah kebutuhan pembangunan masyarakat Mamberamo dan keberlanjutan ekologis dari ekosistem yang rentan ini. Visi kami saat ini antara lain adalah mengkaji pengelolaan daerah aliran sungai yang luas, yang akan dilaksanakan oleh pemerintah, lembaga non-pemerintah (LSM), dan perwakilan masyarakat adat. Hal itu akan dibuat dengan mempertimbangkan kajian sosial dan ekonomi yang akan dikembangkan oleh CI-Indonesia melalui analisa Ekonomi Konservasi dan Sumberdaya. Tujuannya adalah menghasilkan rencana pemanfaatan sumberdaya yang akan memenuhi kebutuhan ekonomi masyarakat lokal Mamberamo, memberikan peluang untuk mengkonservasi hutan dan keragaman hayatinya, dan memungkinkan pembangunan tingkat propinsi yang sesuai sehingga tidak mengancam nilai-nilai alam yang kritis di Mamberamo. Dari sudut pandang konservasi, seluruh aktivitas tersebut akan mengarah pada pembentukan Koridor Mamberamo Raya, yang menghubungkan berbagai sumberdaya alam utama di Mamberamo ke dalam sebuah jaringan bentang alam yang dilindungi, yang akan mengkonservasi daerah aliran sungai, kehidupan masyarakat lokal, dan keragaman hayatinya. Untuk menyelesaikan program-program konservasi yang kompleks itu, kami berharap dapat melakukan serangkaian penelitian tambahan, termasuk survei biologi dan pengembangan kesadaran lingkungan. Hasil-hasil yang diperoleh dari kegiatan pelatihan dan survei RAP ini, telah membuat para pihak di Mamberamo tertarik dan bangga, sehingga sekarang mereka lebih memahami betapa luar biasanya hutan mereka.

Ekspedisi–ekspedisi terdahulu ke wilayah Dabra

Orang-orang Eropa yang pertama mencapai daerah Dabra adalah bagian dari Ekspedisi Opperman, sebuah survai militer Belanda yang tujuan utamanya mengamati Pegunungan Van Rees dan pulau–pulau pesisir di sekitar muara Sungai Mamberamo. Pada bulan November 1914, sebuah detasemen dari ekspedisi ini yang dipimpin oleh L.A.C.M. Doorman, melakukan perjalanan ke hulu Sungai Mamberamo dalam usaha mencapai pegunungan tengah. Berbelok ke timur menuju Sungai Idenburg, mereka menyusurinya hingga mencapai titik terdekat dengan kaki bukit–kaki bukit terpencil. Di sini mereka mendaki, mengikuti sungai jernih yang masuk dari tepi sungai bagian selatan, mencapai hulu/kepala air pada puncak gunung dengan ketinggian 3580 m. Sungai tersebut kemudian diberi nama Sungai Doorman dan gunung di bagian hulu dinamakan Puncak Doorman.

Pada tahun 1920 sebuah ekspedisi besar dari Belanda, dengan 800 orang dipimpin Kapten J.H.G. Kremer, kembali ke lokasi yang sama untuk menjelajah lebih jauh lagi di pegunungan tengah. Kelompok besar ini datang dengan menggunakan perahu melalui rute yang sekarang sering digunakan di Sungai Mamberamo dan Idenburg. Mereka membangun *camp* yang dinamakan Prauwen – bivak, di sebelah hulu dari muara sungai Doorman pada Idenburg, yang berdekatan dengan Dabra. Beberapa kelompok dari ekspedisi ini mendaki Puncak Doorman sekali lagi, lalu dilanjutkan ke arah selatan menuju lembah Sungai Swart – yang mengalir ke Sungai Rouffaer –, cabang bagian barat yang paling besar dari sistem Sungai Mamberamo. Dari sini mereka melanjutkan ke arah selatan melewati dataran tinggi yang tidak rata ke hulu-hulu Sungai Baliem. Walaupun kelompok – kelompok sebelumnya pada akhirnya mencapai Puncak Trikora (pada saat itu dikenal sebagai Puncak Wilhelmina), seluruh ekspedisi tidak menemukan Lembah Baliem - yang letaknya di sebelah timur rute ekspedisi mereka. Meskipun sudah menghabiskan hampir satu setengah tahun di daerah Dabra dan pegunungan di sekitarnya, ekspedisi besar Kremer secara mengagumkan menghasilkan sejumlah hasil-hasil ilmiah yang penting, selain laporan botani, termasuk peta wilayah, oleh Lam (1945). Dengan demikian, survai RAP kali ini tetap merupakan ekspedisi pertama untuk membuat koleksi-koleksi biologi yang sistematik dari daerah ini.

Ekspedisi Archbold III (Archbold *dkk.*, 1942) melakukan pengamatan udara di atas Sungai Doorman dan sekitarnya pada akhir bulan Juni 1938 dengan pesawat apung PBY bernama "Guba". Tujuannya adalah untuk menentukan apakah sungai tersebut memungkinkan menjadi rute untuk memasuki *pegunungan* yang lebih tinggi. Mereka menggambarkan

kawasan tersebut sebagai " … wilayah yang ganas, negeri terlarang, suksesi dari lembah-lembah gelap yang dalam dan punggung-punggung gunung yang berhutan, yang masing-masing tingginya mencapai beberapa ribu kaki …". Namun demikian, ekspedisi ini memilih lokasi perkemahan lebih jauh ke arah timur di Sungai Idenburg, sebuah lokasi yang mereka sebut *Camp*/Perkemahan Bernhard, sekitar 60 km ke arah hulu dari Prauwen-bivak yang dibangun oleh Ekspedisi Kremer. Penduduk lokal di Dabra menyatakan masih mengetahui lokasi *Camp* Bernhard, yang dapat ditempuh dengan perahu motor selama 8 jam. *Camp* Bernhard terletak pada ketinggian 50 meter di jurang tebing yang menghadap ke kumpulan air sungai, dan sekitar 9 meter di atas air ketika dibangun pada akhir Juni 1938. Pada Maret 1939, saat puncak musim hujan (Tabel 1.1) perkemahan ini seluruhnya tenggelam dengan kedalaman 0,3 m, walaupun pada dataran tertinggi. Kejadian ini menunjukkan adanya fluktuasi yang besar ketika musim hujan dan sangat mempengaruhi tinggi air sungai di bagian ini dari daerah aliran Sungai Mamberamo.

Dari camp ini, ekspedi menyusuri bagian pegunungan yang berhadapan dengan Meervlakte, mendirikan serangkaian kemah–kemah untuk mengumpulkan data-data ilmiah yang lebih banyak (Tabel 1.2), yang berakhir di Top Camp pada ketinggian 2150 m. Ekspedisi ini tidak dilanjutkan ke dataran lembah Sungai Hablifuri melalui daerah yang yang topografinya semakin tidak rata. Ekspedisi ini justru menggunakan pesawat apung untuk mengantar-jemput anggota-anggota ekspedisi dari *camp* Bernhard ke sebuah *camp* di Danau Habbema dekat dengan dasar Puncak Trikora. Dari sini, penelitian tumbuh-tumbuhan dan suku-suku di lokasi yang lebih tinggi dapat dilakukan dengan mudah. Selama masa transit ini, ekspedisi menemukan suku dataran tinggi Dani yang mengagumkan dari Lembah Baliem, dan juga sebuah danau di ketinggian menengah di bagian atas dari

Tabel 1.1. Data curah hujan di Bernhard Camp, daerah aliran Sungai Mamberamo, Papua, 1938 – 1939 (Archbold *dkk.*, 1942).

Bulan	Curah Hujan (mm)
Juli 1938 (21 hari)	100
Agustus 1938	259
September 1938	718
Oktober 1938	304
Nopember 1938	547
Desember 1938	454
Januari 1939	561
Februari 1939	511
Maret 1939	539
April 1939	802
Total (10 bulan)	**4795**

Tabel 1.2. Lokasi-lokasi Camp Ekspedisi Archbold 1938–1939 (Archbold *dkk.*, 1942).

Nama Camp/Perkemahan	Ketinggian (m)	Perkiraan Titik Koordinat
Bernhard Camp	50	3°29' LS, 139°13' BT
Araucaria Camp	850	3°30' LS, 139°11' BT
Rotan Camp	1200	3°30' LS, 139°09' BT
Mist Camp	1800	3°30' LS, 139°05' BT
Top Camp	2150	3°30' LS, 139°02' BT
Bele River Camp (Ibèlè Camp)	2200	N/A
Mosbosch Camp	2800	N/A
Brievenbus Camp	3560	4°13' LS, 138°44' BT
Puindal Camp	3800	N/A

aliran sungai Hablifuri. Danau kedua ini dinamakan Danau Archbold, digunakan sebagai lokasi pendaratan pesawat apung, bukan untuk basis aktivitas-aktivitas koleksi ilmiah. Dari sudut cakupan, pengorganisasian dan terbitan-terbitan publikasi ilmiah yang muncul kemudian, Ekspedisi Archbold III merupakan eksplorasi ilmiah paling signifikan yang pernah dilakukan di Daerah Aliran Sungai Mamberamo. Hasil-hasilnya, terutama dalam hal koleksi-koleksi flora dan fauna dari *camp* Bernhard dan kaki-kaki bukit di sekitarnya, merupakan landasan yang mengagumkan sehingga apa yang diperoleh dari hasil RAP dapat dibandingkan.

Setelah Ekspedisi Archbold III dan Perang Dunia-II, hanya sedikit pekerjaan ilmiah tambahan yang dilakukan di wilayah Dabra. Pada tahun 1995, perusahaan pertambangan PT. Freeport Indonesia melakukan kegiatan eksplorasi mineral di daerah tangkapan air Sungai Doorman dan sekitarnya. Dabra merupakan basis untuk mensuplai kebutuhan eksplorasi yang diangkut dengan menggunakan helikopter. Tidak dilakukan koleksi biologi selama survai ini, dan tidak dilakukan kegiatan penambangan lanjutan di wilayah ini.

Dari sudut sejarah ilmiah, Dabra jelas merupakan salah satu wilayah yang lebih aktif di bagian utara tengah New Guinea, dengan sejumlah ekspedisi yang melewati daerah ini untuk menuju *pegunungan* di sebelah selatan Sungai Idenburg. Sayangnya, rangkaian ekspedisi-ekspedisi ini tidak menghasilkan akumulasi koleksi spesimen-spesimen biologi yang sepadan. Sedikit sekali informasi sejarah alam mengenai daerah Dabra sebelum dilakukannya ekspedisi RAP, kecuali koleksi-koleksi yang dihasilkan oleh Ekspedisi Archbold III dari *camp* Bernhard dan sekitarnya ke arah Timur.

Berdasarkan laporan Archbold (Archbold *dkk.*, 1942) keberadaan penduduk asli lembah Sungai Mamberamo pada saat kontak pertama dilakukan adalah "Meervlakte dan sekitarnya dihuni oleh orang-orang yang hidup berpindah-pindah yang tinggal berpencar-pencar … mereka menyebut dirinya Tabbertoea. Pada waktu banjir, mereka naik ke perahu beserta anjing-anjingnya dan beberapa keperluan rumah tangga dan menghilang dari lingkungan tersebut. Mereka kembali ketika air sudah mulai surut. Gambaran Van

Arcken's mengenai pria suku ini adalah tinggi besar, terlihat liar, tubuhnya tertutup tebal dengan kurap/panu yang gatal. Tampaknya mereka tidak menanam apapun kecuali kalau mereka yang mananam buah sukun yang banyak tumbuh di sepanjang tepian sungai. Mereka hanya membangun rumah-rumah sementara, memenuhi kebutuhan sehari-hari dari sagu, berburu dan menangkap ikan, dan kemungkinan terkadang pergi jauh dari perahunya yang aneh" (perahu berupa batang kayu yang dilubangi).

Perahu-perahu penduduk asli juga menarik perhatian Lam (1945), yang melewati wilayah ini tahun 1920 dan memberi komentar "…perahu-perahu yang unik seperti kita lihat hanya ada pada Meervlakte. Dibuat dari batang-batang kayu berat, bagian bawahnya lebar, bagian badannya agak dalam, sisi-sisinya melengkung ke dalam, terdapat celah sempit yang bagian atasnya sempit. Bagian haluan dan buritan dibuat melintang, sehingga dilihat dari depan tampak seperti tumpul, dilihat dari atas seperti sempit. Bentuk kano dibuat sedemikian rupa, mungkin disebabkan oleh sikap tidak bersahabat dari suku-suku berbeda yang hidup di sepanjang sungai. Orang-orang yang berada di kano dapat bersembunyi dengan baik di balik sisi-sisi perahu, sambil tetap bisa mengawasi lawan-lawannya melalui lubang-lubang".

Perahu-perahu tersebut saat ini tampaknya menjadi barang kuno di Sungai Mamberamo dan anak-anak sungainya karena yang terlihat pada waktu survai RAP ini adalah jenis perahu-perahu bersisi rendah yang khas di New Guinea modern. Selain itu, kebanyakan penduduk wilayah ini -- yang jumlahnya masih sangat sedikit--, terkonsentrasi di sekitar kota-kota permanen yang memiliki landasan pesawat, sepert Dabra. Perubahan yang baru terjadi ini bisa diketahui dari tidak adanya pemukiman di Dabra pada waktu Ekspedisi Archbold III melewati daerah ini antara 1938 dan 1939.

Gambaran Lokasi-lokasi RAP di Dabra

Wilayah Dabra terdiri atas topografi yang curam dan transisi geologi dari dataran, dataran berawa-rawa di pedalaman daerah aliran Sungai Memberamo, ke lereng-lereng terjal di bagian utara pegunungan tengah New Guinea. Dataran rendah Memberamo terdiri dari endapan-endapan aluvial kuarter yang tererosi dari sisi-sisi pegunungan di sekitarnya, --di mana Dabra mewakili sisi utara dari kompleks busur pulau Eosin tambahan yang menunjukkan lajur-lajur yang berurutan dari utara ke selatan dimulai dari masa pertengahan ke masa akhir vulkanis tersier, opfiolit Eosin dan metamorfis Eosin (Dow *dkk.*, 1986). Lokasi studi untuk RAP ini terutama berada di jalur vulkanis, yang umur bebatuannya belum diketahui dengan akurat tetapi mungkin dari masa Eosin, karena hubungannya dengan sisa kompleks busur. Terdapat bukti adanya kesamaan antara vulkanik Dabra dan yang terlihat di Sungai Logari yang disurvei pada RAP Wapoga tahun 1998 (Mack dan Alonso, 2000), terutama dengan keberadaan material *(hornfels)* berwarna hitam yang khas di dasar sungai. Lokasi Sungai Furu memiliki tipe campuran batu yang kurang beragam dibandingkan Sungai Tiri; Substrat dasar sungai Furu seluruhnya berasal dari bahan-bahan vulkanik, sedangkan di Tiri materinya terdiri dari contoh-contoh bebatuan metamorphis dan ofiolitik, termasuk marmer. Hal itu mengindikasikan bahwa hulu-hulu sungai masuk jauh ke dalam pegunungan, yang merupakan bagian pinggir dari jalur metamorfis.

Transisi geologi Dabra saat ini juga secara ekologis merefleksikan distribusi tipe-tipe hutan dan habitat-habitat perairan yang terkait dengannya. Endapan aluvial dari jaman Kuarter dari dataran rendah Memberamo ditutupi oleh hutan-hutan rawa yang luas yang berhubungan dengan danau-danau dan saluran-saluran air, mirip dalam banyak hal dengan contoh endapan yang diambil dari bagian hilir Sungai Wapoga, di bawah Siewa pada tahun 1998. Di Siewa sendiri, terdapat zona luas hutan dataran rendah "terre firme", terletak antara hutan-hutan rawa dan tepi kaki gunung. Di Dabra, transisi dari rawa ke kaki bukit sangat drastis, dengan hutan dataran rendah "terre firme" yang hanya berada di zona sempit di dasar perbukitan, atau di bagian yang agak tinggi seperti di Sungai Doorman. Sungai Tiri terdapat pada daerah yang disebutkan terakhir, memiliki dasar sungai yang berpasir dan berkerikil, sama dengan yang terlihat di daerah Siewa sewaktu RAP 1998. Kebalikannya, Sungai Furu memiliki transisi yang cepat ke dasar yang berbatu dan curam, mirip dengan sungai-sungai di daerah Logari. Pendeknya zona habitat perairan di sepanjang Sungai Furu merefleksikan bahwa di daerah Logari, jalur vulkanis dari lereng kaki perbukitan sebelah luar yang curam menjadi dataran—dataran alluvial berawa, dengan sedikit sedimen dimana hutan-hutan dataran rendah dan sungai berasal. Diskusi lebih jauh mengenai pola-pola zonasi hutan di sepanjang Memberamo dapat ditemukan di Archbold *dkk.* (1942).

Status Hukum dan Upaya Konservasi Saat ini

Di daerah Aliran Sungai Mamberamo, Pemerintah Indonesia telah menetapkan Suaka Margasatwa Sungai Rouffaer (310,000 ha) pada tahun 1980 and Suaka Margasatwa Pegunungan Foja (1,018,000 hectares) pada tahun 1982. Terletak di tengah-tengah kawasan Mamberamo dengan ketinggian mulai dari permukaan laut hingga 2.100 meter. Penentuan kedua Suaka Margasatwa tersebut bertujuan melindungi perwakilan ekosistem dataran rendah sebelah utara Papua yang berbeda, flora dan fauna Pegunungan Foja, dan tempat bersarang Buaya New Guinea yang terancam, *Crocodylus novaeguinea*. Kawasan itu juga untuk melindungi lebih dari 150 spesies burung, termasuk lima spesies burung Cenderawasih dan Cikukua Mamberamo, *Philemon brassi*.

Walaupun kedua suaka tersebut secara resmi telah ditentukan dan diumumkan, belum ada aktivitas pengelolaan di lapangan. Keduanya belum mempunyai rencana pengelolaan atau infrastruktur pengelolaan yang memadai. Dalam National Conservation Plan for Indonesia (Ministry of Forestry, 1995), kedua daerah itu direkomendasi untuk digabung dan diperluas, kemudian ditetapkan sebagai Taman Nasional dengan luas total 1,749,200 hectares.

Sebagian besar masyarakat Mamberamo masih melakukan perburuan subsisten, mencari ikan di Sungai Mamberamo, dan pertanian skala kecil. Ada juga masyarakat yang memperoleh pendapatan dari berburu buaya dan mengumpulkan gaharu. Secara umum, masyarakat lokal masih sangat bergantung pada lahan mereka dan sumberdaya di sekitarnya.

Bekerjasama dengan mitra lokal, Conservation International ingin mendemonstrasikan pendekatan yang menyeluruh (multi dimensi) untuk memelihara keutuhan keragaman hayati di daerah aliran sungai Mamberamo. Pendekatan ini berupaya menyeimbangkan pengembangan ekonomi dengan pemanfaatan yang berkelanjutan oleh masyarakat. CI saat ini bekerjasama dengan kelompok masyarakat lokal, Dewan Masyarakat Adat Mamberamo, yang ingin melindungi tanah mereka dari industri-industri ekstraktif dan untuk mengembangkan hutan lindung masyarakat yang akan mencakup beberapa zona untuk pemanfaatan, perburuan, budaya, serta perlindungan terhadap daerah inti dan keragaman hayatinya.

DAFTAR PUSTAKA

Archbold, R., A.L.Rand, and L.J.Brass. 1942. Results of Archbold Expeditions. No.41. Summary of the 1938-1939 New Guinea Expedition. Bull. Am. Mus. Nat. Hist. 79: 197–288.

Cheesman, L.E. 1938. The Cylops Mountains of Dutch New Guinea. Geog. J. Lond. 91:21–30.

Davies, H.L. 1990. Structure and evolution of the border region of Papua New Guinea. *In*: Petroleum Exploration in Papua New Guinea: Proceedings of the First PNG Petroleum Convention, Port Moresby, 12–14th February 1990. G.J. Carman and Z.Carman (eds.). Papua New Guinea: PNG Chamber of Mines and Petroleum.

Dow, D.B., G.P. Robinson, U. Hartono, and N. Ratman. 1986. Geologic map of Irian Jaya, Indonesia. Geological research and Development Center, Ministry of Mines and Energy, Bandung. 1:1,000,000 scale map.

Flannery, T.F. and C.P. Groves. 1998. A revision of the genus *Zaglossus* (Monotremata, Tachyglossidae), with description of new species and subspecies. Mammalia 62: 367–396.

Kroenke, L.W. 1984. Cenozoic development of the South-west Pacific. United Nations Economic and Social Commision for Asia and the Pacific, Committee for Co-ordination of Joint Prospecting for Mineral Resources in South Pacific Offshore Areas, Technical Bulletin 6.

Lam, H.J. 1945. Fragmenta Papuana (Observations of a naturalist in Netherlands New Guinea). Sargentia, 5: 1–96.

Ministry of Forestry. 1995. National Conservation Plan for Indonesia: Vol. 9. Irian Jaya Biogeographic Region and Province, Ministry of Forestry, Jakarta.

Pigram, C.J. and H.L. Davies. 1987. Terranes and the accre-tion history of the New Guinea orogen. Bureau of Min-eral Resources, J. Aust. Geol. Geophys. 10: 193–211.

Polhemus, D. and J.T. Polhemus. 1998. Assembling New Guinea: 40 million years of island arc accretion as indicated by the distributions of aquatic Heteroptera (Insecta). *In*: Hall, R. and J.D. Holloway (eds.). Bioge-ography and geological evolution of SE Asia. Leiden: Backhuys. Pp. 327–340.

Ratcliffe, J.B. 1984. Cagar Alam Pegunungan Cyclops Irian Jaya: Management Plan 1985-1989. Jayapura: World Wildlife Fund.

van Royen, P. 1965. Sertulum Papuanum 14. An outline of the Flora and vegetation of the Cyclops Mountains. Nova Guinea N.S. Botany. 21: 451–469.

Chapter 2

Plant diversity in lowland forests of the Yongsu area, Papua, Indonesia

Yance de Fretes, Conny Kameubun, Ismail A. Rachman, Julius D. Nugroho, Elisa Wally, Herman Remetwa, Marthen Kabiay, Ketut G. Suartana, and Basa T. Rumahorbo

CHAPTER SUMMARY

- This study examined plant diversity and forest composition in lowland rainforest at Yongsu, Papua, using five 0.1 ha modified Whittaker plots.

- 197 species of plants were recorded during the RAP survey.

- 293 trees (≥10 cm dbh) representing 88 species in 33 families were recorded in the plots. In addition, 27 species of poles (smaller trees, 5–9.9 cm dbh) in 19 families, 32 species of saplings (1–4.9 cm dbh) in 23 families, and 50 species of grasses and seedlings in 34 families occurred within the plots.

- The canopy is dominated by *Manilkara fasciculata*, *Mastixiodendron pachyclados*, *Palaquium ridleyi*, and *Parastemon urophyllus*.

- Floral richness in lowland forest at Yongsu is similar to other tropical regions in the Asia-Pacific region but is lower than in Latin America.

RINGKASAN BAB – TUMBUH-TUMBUHAN YONGSU

- Penelitian ini mempelajari komposisi hutan dan keanekaragaman tumbuhan di hutan hujan dataran rendah Yongsu, Papua, dengan menggunakan 0,5 hektar *Modified Whittaker Plots*.

- Jumlah total yang diperoleh selama survei adalah 197 spesies tumbuhan.

- 293 pohon (≥ 10 cm dbh) mewakili 88 spesies dari 33 famili berhasil dicatat olalem plot. Selain itu, 27 spesies tingkat pancang (pohonkecil 5–9, 9 cm dbh) dalam 19 famili, 32 spesies tingkat tiang (1–4, 9 cm dbh) dalam 25 famili, dan 50 spesies rumput-rumputan dan semai dari 34 famili dapat ditemukan daleum plot.

- Tingkatan kanopi didominasi oleh *Manilkara fasciculata*, *Mastixiodendron pachyclados*, *Palaquium ridleyi* dan *Parastemon urophyllus*.

- Kekayaan jenis tumbuh-tumbuhan di hutan dataran rendah Yongsu mirip dengan kawasan tropis lainnya di Asia-Pasifik tetapi lebih rendah dari Amerika Latin.

INTRODUCTION

Tropical forest habitats are being fragmented and destroyed at an alarming rate. Forest loss in Indonesia alone is estimated to be about 1.7 million ha per year (World Bank, 2001). Without radical forest conservation measures most lowland rainforest in Sumatra and Kalimantan will be gone by 2005 and 2010 respectively (World Bank, 2001). Deforestation rates are lower in Papua but future logging and mega-projects such as the Mamberamo dam have the potential to devastate large areas of forest. Papua's forests contain an exceptional diversity of spectacular and endemic plant and animal species, representing nearly half of the total biodiversity of Indonesia (Conservation International, 1999).

Forests in the Jayapura region, including the slopes of the Cyclops Mountains, have been severely impacted by expansion of human settlements and agricultural activity. Periodic flooding and water shortages in Jayapura reflect damage to the surrounding forest ecosystems, which form significant water catchments for the city's water supplies. Despite their proximity to Jayapura no meaningful efforts have been made to protect these forests, and the fauna and flora of the Cyclops Mountains Nature Reserve remains poorly documented and under threat.

The objective of this study was to document plant species richness and composition in Jari forest on the northern slopes and coastal fringe of the Cyclops mountains west of Jayapura, and to compare these results with similar studies undertaken in other tropical regions.

METHODS

Study area

This study was conducted in Jari forest (02°26'15"S, 140°30"E) near Yongsu Dosoyo (Map 1) on the northern fringe of the Cyclops Mountains. The study site is approximately 100 m a.s.l. in relatively flat lowland rain forest growing on utisol soil. During the study average soil pH was 6.5, mean soil temperature was 27°C, and mean relative humidity beneath the canopy was 75%. Mean monthly rainfall during five years at a climate station near Yongsu Dosoyo was 666.9 mm with the heaviest mean rainfall recorded in April (1075.2 mm) and the lowest mean rainfall recorded in September (256.6 mm).

Human activities in the Jari forest appear to be restricted to low-intensity hunting by local people. A narrow trail through the forest is used occasionally as an alternative route to neighbouring Ormu Village, especially when rough seas make the ocean crossing unsafe.

Sampling procedures

Plants were sampled in five 20 x 50 m (0.1 ha) modified Whittaker plots (Figure 2.1; Stohlgren, 1995) that were placed randomly in the forest within a 1 km radius of our

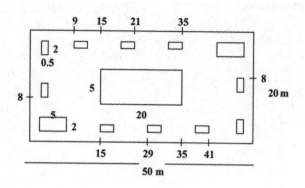

Figure 2.1. Design of Whittaker modified plots used in vegetation analyses. Five Whittaker plots were used to sample vegetation at Jari (Yongsu) and Furu (Mamberamo). All distances are in meters.

makeshift camp. Plants were categorized into four size classes:

1) Trees (≥ 10 cm dbh),

2) Poles (smaller trees 5–9.9 cm dbh),

3) Saplings (1–4.9 cm dbh), and

4) Seedlings and grasses (< 50 cm tall).

Plots were divided into smaller subplots (Figure 2.1; Stohlgren, 1995; Conservation International, 2000):

*   10 subplots of 0.5 x 2.0 m were located at the periphery of the plot to sample herbaceous saplings, seedlings, and grasses;

*   2 subplots of 2 x 5 m were located at opposite corners of the plot to sample trees and saplings;

*   1 subplot of 5 x 20 m was located in the center to sample poles and trees.

Thus the area sampled within the total 0.5 ha of plots was different for each plant category. All plants within the plots were identified and counted.

Observations

Additional surveys were conducted around the campsite, along forest trails and the beach front, and in village gardens and secondary forest. Fresh voucher specimens were soaked in 70–80% ethanol to allow more detailed future examination and are stored at Lembaga Biologi Nasional-LIPI (LBN-LIPI) Bogor.

Data analysis

Our analyses and discussion will focus largely on trees (≥10 cm dbh) because of substantial difficulties encountered when trying to identify the smaller grasses, seedlings, and saplings. Furthermore, we will focus primarily on data gathered from Whittaker plots. To determine whether Whittaker plots adequately sampled plant diversity we pooled data from plots and subplots to construct species-area curves. To examine species similarity between plots we used Krebs's (1989) computer program to calculate the Morisita Index of Similarity. Whenever appropriate our data are compared to previous studies in tropical lowland forests (e.g. Kameubun, 2000).

RESULTS

Species composition and richness

The forest at Jari is lowland primary rainforest. Tree density ranged from 46–76 stems in 0.1 ha. The largest tree was *Manikara fascicucata* (Sapotaceae) with a maximum diameter of 121 cm. Other large species including *Campnosperma brevipetiolata* (Anacardiaceae) and *Mastixiodendron pachyclados* (Rubiaceae) reached 80 cm in diameter. Forest composition is not the same in the different size classes (Figure 2.2). Lists of plant species recorded in Jari Forest plots are presented in Appendices 1–4. Major results from the Whittaker plots include: 50 species of grasses, herbs, and seedlings in 34 families; 32 species of saplings in 23 families; 27 species of poles in 19 families; and 88 species of trees in 33 families. The vegetation is substantially different from that documented in the Mamberamo Basin (de Fretes et al., this volume; Appendix 9). The family Myrtaceae is a dominant component of the forest flora based on number of species alone, while *Syzygium* sp. and *Tejsmanniodendron bogoriense* were the most common species in the five Whittaker plots.

Emergent canopy is formed by *Manilkara fasciculata*, *Mastixiodendron pachyclados*, *Palaquium ridleyi*, *Parastemon urophyllus*, *Parastemon verstegii*, *Campnosperma brevipetiolata*, and *Pometia pinnata*. The second canopy layer is dominated by *Gymnacranthera paniculata*, *Horsfieldia helwigii*, *Horsfieldia sylvestris*, *Myristica lancifolia*, *Syzygium* spp., *Canarium* spp., *Pimeleodendron amboinicum*, *Celtis philippinensis*, *Gnetum gnemon*, *Teijsmanniodendron ahernianum*, *Teijsmanniodendron bogoriense*, *Tetractomia obovata*, *Hopea novoguinensis*, and *Diospyros discolor*. The third canopy layer is dominated by *Garcinia dulcis*, *Garcinia celebica*, *Gonocaryum littorale*, *Gonocaryum* sp., *Medusanthera laxiflora*, *Gomphandra montana*, *Dysoxylum* sp., and *Timonius novoguinensis*. The forest floor is dominated by *Trichomanes* spp., *Selaginella* spp., *Mapania* sp., *Paramapania parvibracteata*, *Dianella* sp., *Elatostema* sp. and *Carex* sp.

Woody lianas include *Tetracera* sp., *Tetrastigma* sp., *Rhaphidophora novoguinensis*, *Rhapidophora verstegii*, *Rhaphidophora* sp., *Pothos scandens*, *Halochlamis beccarii*, *Tecomanthe cyclopensis*, *Freycinetia angustifolia*, *Freycinetia linearis*, *Freycinetia excelsa*, *Strychnos ignatii*, and *Mucuna* spp. Commonly observed epiphytes include *Dendrobium* spp., *Agrostophyllum* sp., *Asplenium nidus*, *Asplenium* sp., *Drynaria* sp., and *Myrmecodia lanceolata*.

Species similarity

The Morisita Index of Similarity for trees among plots is presented in Table 2.1. Values close to 1 indicate high similarity (reflecting low diversity) between plots, while values close to 0 indicate low similarity (high diversity). Our results indicate that there were high species similarities among plots 1–2, 3–4, and 1–5. Plot 2 is similar to plot 1, but very different from other plots. Similarity index values for comparisons between Jari and Furu (0.23) and Tiri (0.19) are extremely low (de Fretes et al., this volume).

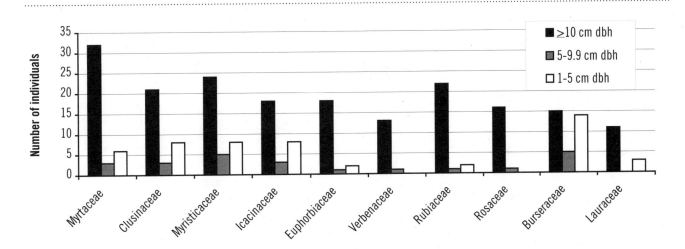

Figure 2.2. Abundance of trees in three size classes for the 10 dominant families at Jari forest.

Table 2.1. Morisita Index of Similarity between plots in Jari forests (all trees ≥ 5cm were used in this calculation).

	Plot 2	Plot 3	Plot 4	Plot 5
Plot 1	0.92	0.68	0.69	0.74
Plot 2		0.56	0.39	0.35
Plot 3			0.79	0.59
Plot 4				0.61
Plot 5				

Species area curves

Species area curves for all plant categories indicate that sampling effort was adequate to document vegetation diversity only for trees (Figure 2.3). Additional sampling will undoubtedly substantially increase the number of species recorded from Jari forest.

General vegetation surveys

Tree species commonly observed in secondary forests or old gardens around Yongsu Dosoyo were *Macaranga aleuritoides, Alphitonia incana, Fagraea racemosa, Eoudia elleryana, Cananga odorata, Endospermum peltatum, Ficus damaropsis, Campnosperma auriculata,* and *Macaranga* sp. At least 68 species in 41 families were observed in secondary forests and along beaches. Beach vegetation included 38 plant species in

A. Grasses, herbs, and seedlings (< 50 cm in height).

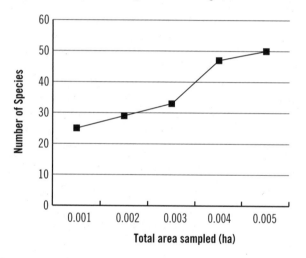

B. Saplings (diameter 1–4.9 cm dbh).

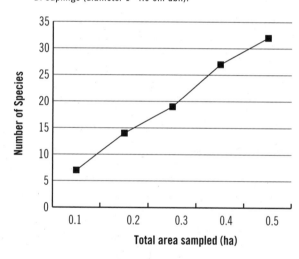

C. Poles (diameter 5–9.9 cm dbh).

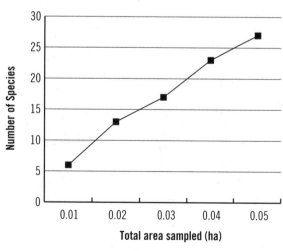

D. Trees (diameter ≥ 10 cm dbh).

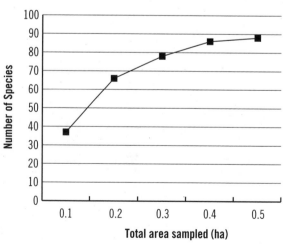

Figure 2.3. Species accumulation curves for four different size categories of plants in Jari Forest, Yongsu.

25 families, with the family Sterculiaceae dominating the vegetation (Appendix 5). About 30 plant species in 23 families were recorded in secondary forests (Appendix 6). A total of 178 plant species in 56 families were recorded in general surveys in Jari forest and around Yongsu Dosoyo (Appendix 7).

DISCUSSION

Our species-area curves based on plots indicate that some plant categories were reaching an asymptote after sampling area had reached 0.5 ha. However, the high number of additional tree species documented outside of plots indicate that more sampling is required before the majority of species in the Yongsu area have been documented. This conclusion is supported by a comparison of canopy composition at Jari with a recent study at nearby Yongsu Dosoyo (Kaemeubun, 2000). While canopy species at Jari include *Manilkara fasciculata, Mastixiodendron pachyclados, Palaquium ridleyi, Parastemon urophyllus, Parastemon verstegii, Campnosperma brevipetiolata,* and *Pometia pinnata,* at Yongsu the canopy flora is dominated by *Schima wallichi, Elaeocarpus glaber, Pometia pinnata, Gymnacranthera paniculata, Palaquium amboinense, Hopea novoguinensis,* and *Schizomeria serrata* (Kaemeubun, 2000). All of these data indicate that tree diversity in this region is unlikely to be adequately documented until at least 1 ha (or 10 Whittaker plots) has been sampled. Although comparisons of tree diversity across tropical regions are frequently compromised by the use of different methods for quantifying diversity, it does appear that, in general, tree diversity at Jari is similar to that reported elsewhere in the tropics (Table 2.2; Whitmore, 1975; McDade et al., 1994).

The relative abundances of *Canarium* sp., *Diospyros discolor, Elaeocarpus* sp., *Syzygium* sp., and *Palaquium* sp. in the seedling category differed from their relative abundances at the pole or tree stages. A long-term study using permanent plots should be undertaken to further quantify changes in species composition with changes in age structure. Documentation of the long-term dynamics of this forest assemblage should be feasible given the proximity of the site to Jayapura, the lack of foreseeable threats to the forest, and the support for conservation projects received from the Yongsu community during this study.

Jari forest has an alpha diversity (species richness) greater than that quantified from two study plots at Yongsu proper (Kaemeubun, 2000) and the species similarity index reveals a lower similarity between plots at Jari than those at Yongsu. However, careful mapping of plots should be undertaken so that species similarity data can be critically assessed to remove confounding factors such as distance between plots. This information will also assist in determining the distance between plots at which maximum information about plant diversity can be obtained.

Forests around Yongsu provide valuable material for house and canoe construction by local villagers, but extraction of

Table 2.2. Number of trees (≥10 cm dbh) in other tropical regions. Data from Whitmore (1975) and McDade at al. (1994).

Location	Plot Area (ha)	# species	Years Recorded
Moluccas (Halmahera/ Maluku utara)	0.5	76	1987
Kalimantan	1	149; 129; and 128	1981
Panama (USA)	1.5	130	1994
Papua New Guinea	1	184; 181,152; and 154	1970
Sarawak (Malaysia)	1	214; 223	1983

forest resources appears to be sustainable at this stage and the forests, especially at Jari, are largely intact. Forests around many other villages on the northern edge of the Cyclops Mountains have been severely degraded. The presence of intact lowland forest that is easily accessible from Jayapura makes the Yongsu area an excellent site for biological training and long-term field studies. The Yongsu communities are supportive of conservation programs and training courses undertaken in the region, enhancing the prospects of long-term survival of these forests. Protection of this forest will protect the village from land slides, protect other fauna found in the forest, and ensure that the Yongsu community retains access to a wide range of forest plants that are used locally for medicine or ritual healing (Warpur, 2000).

LITERATURE CITED

Conservation International. 1999. The Irian Jaya Biodiversity Conservation Priority-Setting Workshop: Final report. Washington, DC: Conservation International.

Conservation International. 2000. Teknik survei lapangan program penilaian cepat (Rapid assessment program): kursus pelatihan. Jakarta: Conservation International.

Kamaebun, C. 2000. Keanekaragaman jenis pohon hutan hujan tropis dataran rendah di Desa Yemang, Yongsu Dosoyo, Kecamatan Depapre, Kabupaten Jayapura. Skripsi. Fakultas Keguruan dan Ilmu Pendidikan, Universitas Cenderawasih.

Krebs, C. 1989. Ecological Methodology. New York: Harper Collins.

McDade, L.A, K.S. Bawa, H.A. Hespenheide, and G. Hartshorn. 1994. Ecology and Natural History of a Neotropical rain forest. Chicago: University of Chicago Press.

Stohlgren, T.J. 1995. A modified-Whittaker nested vegetation sampling method. Vegetation. 117: 113–121.

Warpur, M. 2000. Pola Penggunaan Lahan dan Pemanfaatan Sumberdaya Hayati Tumbuhan oleh Masyarakat Yongsu di Sekitar Cagar Alam Pegunungan Cyclops, Irian Jaya. Thesis Program Studi Biologi Program Pascasarjana, Universitas Indonesia.

Whitmore, T. C. 1975. Tropical rain forests of the Far East. Oxford: Clarendon Press.

World Bank. 2001. Indonesia: Environment and Natural Resource Management in a time of transition. Washington, DC: World Bank. (Website version).

Chapter 3

Vegetation of the Dabra area, Mamberamo River Basin, Papua, Indonesia

Yance de Fretes, Ismail A. Rachman, and Elisa Wally

CHAPTER SUMMARY

- This study examined plant richness and species composition in the Furu River and Tiri River areas of the Mamberamo Basin in 0.5 ha of modified Whittaker plots (at Furu) and 0.5 ha of transect lines (at Tiri).

- 281 trees (≥ 10 cm dbh) representing 121 species were recorded in plots at Furu, and 208 trees representing 90 species were recorded in transects at Tiri.

- The canopy is dominated by *Vatica rassak, Hopea novoguinensis, Atuna racemosa, Xanthophyllum tenuipetalum, Ceratopetalum succirubrum, Endiandra euadenia, Madhuca* sp., *Lithocarpus vinkii, Pometia pinnata*, and *Pimeleodendron amboinicum* at Furu, and by *Pometia pinnata, Heritiera littoralis, Microcos ceramensis, Intsia palembanica, Intsia bijuga, Pterigota horsfieldii, Garcinia latissima*, and *Vatica rassak* at Tiri.

- 62% of trees (≥10 cm dbh) at Furu and 54% of trees at Tiri were represented by a single individual.

- Species-area curves indicate that the number of tree species in both study areas is likely substantially higher than we documented.

RINGKASAN BAB – TUMBUH-TUMBUHAN DABRA

- Penelitian ini mempelajari komposisi spesies tumbuhan dan kekayaan jenisnya di Sungai Furu dan Tiri, daerah aliran sungai Mamberamo, dengan menggunakan 0,5 hektar *Modified Whittaker plots* (di Furu) dan 0,5 hektar garis transek (di Tiri).

- 281 pohon (≥ 10 cm dbh) mewakili 121 spesies tercatat di plot Furu, dan 208 pohon mewakili 90 spesies di transek Tiri.

- Tingkat kanopi di Furu didominasi oleh *Vatica rassak, Hopea novoguinensis, Atuna racemosa, Xanthophyllum tenuipetalum, Ceratopetalum succirubrum, Endiandra euadenia, Madhuca* sp., *Lithocarpus vinkii, Pometia pinnata*, dan *Pimeleodendron amboinicum*, sedangkan tingkat kanopi di Tiri didominasi oleh *Pometia pinnata, Heritiera littoralis, Microcos ceramensis, Intsia palembanica, Intsia bijuga, Pterigota horsfieldii, Garcinia latissima*, dan *Vatica rassak*.

- 62% pohon (≥10 cm dbh) di Furu dan 54% pohon di Tiri keberadaannya diwakili oleh satu individu.

- Kurva area-spesies mengindikasikan bahwa jumlah spesies pohon pada dua lokasi penelitian lebih tinggi daripada yang berhasil kami dokumentasikan.

INTRODUCTION

Papua's flora is the richest in Indonesia and among the richest in the world. It also has an extraordinary level of endemism with between 60 and 90% of species known only from New Guinea, many of which are probably found only in Papua (Myers, 1988; Johns, 1995). Papua retains 80% of its forest cover, representing the largest remaining tracts of intact forest in Indonesia. In recognition of the exceptional conservation values of Papua's forests the province has been classified as a Major Tropical Wilderness Area (Myers et al., 2000).

Most information on plant diversity and forest structure in New Guinea comes from field studies in Papua New Guinea (e.g. Paijmans, 1976; Whitmore, 1990). Botanical collections in Papua are scarce with an average of only about 20 specimens collected per 100 square km (Johns, 1995). Collections have been concentrated in the Vogelkop area, Cyclops Mountains, Mount Jaya and Timika areas, and around Mount Trikora and Lake Habbema (Johns, 1995). In a recent unpublished study Kaemebun (2000) documented tree diversity and forest structure in forest around Yongsu, northern Cyclops Mountains, but published data on plant diversity in a "typical 1 hectare plot" are unavailable for forests anywhere in Papua.

In this study we document the composition and diversity of vegetation at two sites in the Dabra area and compare these data with a similar study conducted on the northern side of the Cyclops Mountains (deFretes et al., this volume).

METHODS

Study area

The Mamberamo River Basin covers about 7.7 million hectares of largely pristine and continuous rainforest in northern Papua. The region contains a variety of ecosystems including the northern slopes of the central cordillera, lowland rainforests, extensive wetlands, and estuarine habitats. It is sparsely populated, with an average density of 2 persons per km² and an estimated total population of 7,000 inhabitants.

Sampling procedures

The botanical survey was done at Furu (3°17'S, 133°38'E) between 2–8 September 2000, and Tiri (3°17'S, 138°34'E) betwen 9–14 September (Map 1). Plants were categorized into four size classes:

1) Trees (≥ 10 cm dbh),

2) Poles (smaller trees 5–9.9 cm dbh),

3) Saplings (1–4.9 cm dbh), and

4) Seedlings and grasses (< 50 cm tall).

Field methods were the same as those employed at Yongsu (deFretes et al., this volume) except that at Tiri we sampled two transects (20 x 125 m; Figure 3.1) at randomly selected positions instead of five Whittaker plots (Stohlgren, 1995). The total area sampled (0.5 ha) was the same. In transects trees with diameter ≥ 10 cm were sampled in one subplot of 20 x 25m and five subplots of 20 x 20 m. Saplings (diameter 1.0–9.9 cm; including poles, deFretes et al., this volume) were sampled in 6 subplots of 5 x 5 m located in a zigzag pattern within the plot (Figure 3.1). Grasses, seedlings, and herbaceous plants (≥ 50 cm tall) were sampled in 50

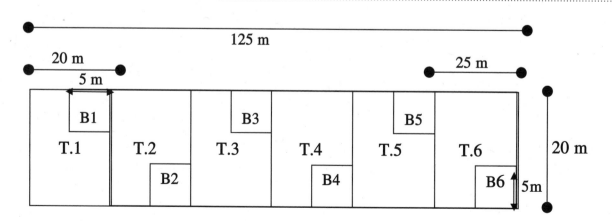

Figure 3.1. Design of transect lines used to sample plants at Tiri (not to scale).

(0.5 x 2m) subplots). All trees were measured, identified, and counted.

Our analyses focus on trees (≥ 10 cm dbh) due to the difficulty of identifying many of the other plant groups. Data from plots (including transects) were used to calculate species richness and total density, and to construct species-area curves. The Morisita Index of Similarity (Krebs, 1989) was used to estimate species similarity between plots at each study site. The similarity index ranges from zero to one where values close to one indicate high species similarity and values close to zero indicate low species similarity.

RESULTS

Furu

The presence of "pioneer species" such as *Macaranga mappa*, *Macaranga* sp., *Nauclea orientalis*, *Endospermum peltatum*, *Mallotus* sp., *Cananga odorata*, *Duabanga moluccana*, and *Glochidion* spp. is evidence that the forest at Furu has suffered low to medium levels of disturbance. The Dabra community uses the area for hunting and collecting forest products, and their main impact appears to be the felling of trees to collect matoa fruit *Pometia* spp. (Sapindaceae). The riverbank is dominated by *Mitragyna speciosa*, *Planchonia* cf. *vallida*, *Leucosyke capitelata*. *Dracontomelon da'o*, *Homalium foetidum*, and *Croton* spp. Mean tree diameter was 21 cm but the largest tree (*Pometia pinnata*) recorded in sample plots had a diameter of 90 cm dbh.

At least 439 saplings and trees representing 161 species were recorded from the plots (Appendix 9), although this figure should be considered preliminary and will probably increase as more botanical vouchers are identified. Of these, roughly 70 species were present as saplings and smaller trees (1.0–9.9 cm dbh) and 121 species as trees with dbh ≥10 cm. Difficulties with identifying grasses, seedlings and other smaller vegetation precluded an estimate of species diversity for these groups.

The emergent canopy is dominated by *Vatica rassak*, *Hopea novoguinensis*, *Atuna racemosa*, *Xanthophyllum tenuipetalum*, *Ceratopetalum succirubrum*, *Endiandra euadenia*, *Madhuca* sp., *Lithocarpus vinkii*, *Pometia pinnata*, and *Pimeleodendron amboinicum*. The second layer is dominated by *Polyalthia discolor*, *Ceratopetalum succirubrum*, *Gnetum gnemon*, *Cryptocarya* spp., *Gymnacranthera paniculata*, *Teijsmanniodendron bogoriense*, and *Myristica* spp., while the under storey includes *Uruphyllum umbeliferum*, *Urophyllum* sp., *Antiriopsis decipiens*, *Pisonia longirostris*, and *Licuala* sp. The forest floor is dominated by *Mapania* spp., *Trichomanes* spp., *Elatostema weilandii*, *Rhaphidophora* sp., and *Adianthum* sp. (Appendix 8). Several lianas (*Freycinetia* sp., *Rhaphidophora* sp. and *Strychnos* sp.) occurred in each of the plots. Emergent canopy reaches up to 50 m, but most trees at Furu and Tiri do not exceed 40 m in height.

Tiri

There were few indicators of forest disturbance at Tiri. The closest disturbed area is a fallow garden located about 2 km from the site. Three hundred and twenty-nine saplings, poles, and trees (≥ 1 cm dbh) representing 128 species were recorded in transect plots, including *Pometia pinnata*, *Chisocheton stellatus*, *Medusanthrea laxiflora*, and *Gomphandra australiana* (Appendix 9). About 64 species were present as saplings and small trees (1.0–9.9 cm dbh), and 90 species as large trees (≥10 cm dbh). The canopy at Tiri is mainly formed by *Pometia pinnata*, *Heritiera littoralis*, *Microcos ceramensis*, *Intsia palembanica*, *Intsia bijuga*, *Pterigota horsfieldii*, *Garcinia latissima*, and *Vatica rassak*. The second layer is dominated by *Pimeleodendron amboinicum*, *Gymnacranthera paniculata*, *Chisocheton stellatus*, *Gnetum gnemon*, *Medusanthera laxiflora*, *Pseusobotrys cauliflora*, *Celtis philippinensis*, and *Teijsmanniodendron bogoriense*. The third layer is dominated by *Rhyticaryum oleraceum*, *Casearia aruensis*, *Pisonia longirostris*, *Maschaloderme simplex*, and *Zygonium calothyrsum*, and the forest floor is mainly covered by *Elatostema*, *Rhaphidophora*, *Tricomanes* spp., *Leptaspis urceolata*, *Mapania*, and *Begonia* spp. Lianas include *Rhaphidophora* sp., *Piper* spp., *Freycinetia* spp., and *Strychnos* sp.

Pometia pinnata, *Terminalia* sp., *Chisocheton ceramicus*, *Aglaia argentea*, *Heritiera littoralis*, *Prunus* sp., *Ficus* sp., *Canarium acutifolium*, *Parkia versteeghii*, *Mastixiodendron pachyclados*, *Octomeles sumatrana*, and *Microcos ceramensis* are common along the riverbank. In disturbed or gap areas, common species include *Macaranga mappa*, *Macaranga* sp., *Bridelia insulana*, *Cananga odorata*, *Endospermum moluccanum*, *Leucosyke capitelata*, *Glochidion* sp., and *Campnosperma auriculata*. Regrowth in the fallow garden is dominated by *Trichospermum pleiostigma*, *Trema* sp., and *Ipomea* sp. The largest trees are *Planchonia papuana* (Lecythidaceae) and *Garcinia latissima* (Clusiaceae), both reaching 80 cm dbh. Mean diameter of trees in plots was 23 cm dbh.

A list of tree species (including saplings) in plots is provided in Appendix 9 and a list of voucher specimens obtained outside plots and transects during the survey is presented in Appendix 10.

Common species and "dominant" families

There was substantial overlap in composition of the 10 most abundant species at Furu and Tiri (Figure 3.2). Exceptions were *Pometia pinnata* that was abundant at Tiri but not at Furu, *Vatica rassak* that was abundant at Furu but not at Tiri, and *Ceratopetalum succirubrum*, which was abundant at Furu but absent from Tiri. The abundance of *Pometia pinnata* at Tiri can be attributed to the greater disturbance of forest at that site.

Similarly, there was little difference between the 10 dominant families (those with the largest number of species) between sites, except for Lauraceae that contained more species at Furu than at Tiri (Figure 3.3). Most species were represented by only one individual. For example at Furu about 60% of saplings were represented by one individual.

The proportions were 60% at Tiri and 54% at Jari. At the tree stage 62% of species at Furu and 52% of species at Tiri were represented by a single plant. At Jari the proportion was much lower (30%), but only three species had more than 10 individuals in 0.5 ha. Table 3.1 shows stem density and species diversity for each size class from Furu (pooled data from five Whittaker plots), Tiri (pooled data from two transect lines), Jari (pooled data from five Whittaker plots), and Yongsu (pooled data from two transect lines; Kameubun, 2000). A species area curve (Figure 3.4) indicates that additional sampling using plot and transect techniques will substantially increase the number of species recorded from the Mamberamo sites.

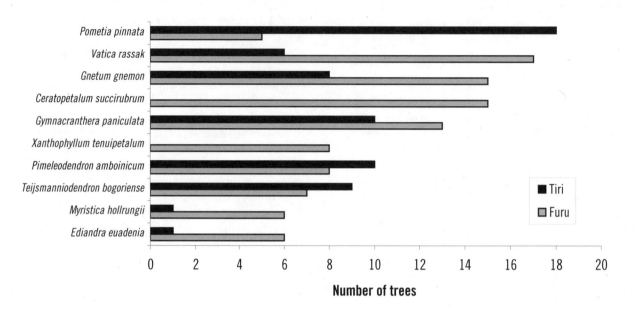

Figure 3.2. Comparative abundance of the ten most common tree species (≥ 10 cm dbh) sampled in five Whittaker plots (20 x 50 m) at Furu and in two transect lines (20 x 125 m) at Tiri.

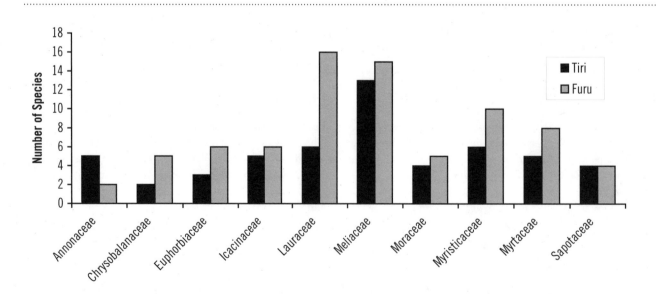

Figure 3.3. Number of species in the ten most common families sampled in five Whittaker plots (20 x 50 m) at Furu and in two transect lines (20 x 125 m) at Tiri.

Species similarity between plots at each site

We calculated similarity indices for saplings, poles, and trees (≥ 1 cm dbh) between five plots at Furu and two transects at Tiri with the Morisita Index of Similarity (Krebs 1989, equation 9.11). Table 3.2 shows that tree diversity indices varied among plots at Furu. Plots 1, 2, and 3 had moderate similarity values. Plots 4 and 5 had the highest similarity value, but these plots had exceptionally low similarity with plots 1–3 (Table 3.2). The two transects at Tiri had a low similarity value of 0.32.

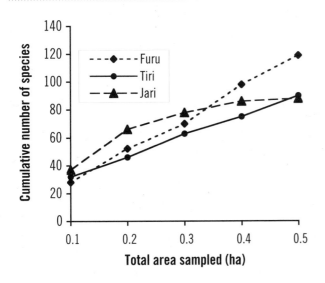

Figure 3.4. Species accumulation curves for trees (≥10 cm dbh) sampled in 5 Whittaker plots (20 x 50 m) at Furu and Jari, and in 2 transect lines (20 x 125 m) at Tiri.

Table 3.1. Stem density and number of tree species at Furu, Tiri, Jari, and two sites near Yemang (Yongsu; 2 sites).

	Furu	Tiri	Jari	Yongsu 1	Yongsu 2
Stem density (0.5 ha)	279	208	293	295	327
No. tree species (≥10 cm dbh)[a]	121	90	88	92	93
No. sapling and pole species (2.0–9.9 cm dbh)[b]	na	41	na	44	48
No. sapling species (1.0–4.9 cm dbh)[c]	50	na	33	na	na

[a] Total area sampled 0.5 ha
[b] Total area sampled 0.03 ha
[c] Total area sampled 0.01 ha

Species similarity between sites

The Furu and Tiri sites had only a moderate similarity index (0.59) for tree species ≥ 10 cm dbh, and both of these sites have a very low similarity index when compared with Jari (Furu vs Jari 0.23, Tiri vs Jari 0.19). These calculations indicate that the three study sites have quite different species compositions, and that the forest at Jari is very different from the Mamberamo sites.

Vegetation surveys

At least 217 additional plant species were recorded during general surveys of vegetation around the two camps and along forest trails, and about 53 additional species (including some introduced species) were documented around Dabra village (Appendix 11). The poor state of knowledge about taxonomy and distribution of Mamberamo vegetation precludes assessment of the conservation status of almost all species recorded in this study. Many of the botanical specimens were previously not represented in the Herbarium Bogoriense, Bogor.

DISCUSSION AND CONCLUSIONS

The number of tree species recorded at Furu was higher than the number recorded at Tiri and substantially higher than the sites we studied at Yongsu (Jari and Yemang; Table 3.1). A similar trend was evident for saplings and smaller trees. Harvesting of large trees at Furu has probably contributed to this increased diversity by creating a mosaic of canopy gaps. Although current harvesting practices are similar to natural disturbances that create gaps, natural events occur randomly and unpredictably over a wider area and time span. At Furu the occurrence of many trees that are typical of disturbed habitats suggests that continuing relatively intensive tree harvesting may be shifting forest composition at this site towards a community dominated by pioneer species.

The low species similarities between plots at each site suggest that plant diversity on a local scale in these forests is quite high. Larger plots placed over a broader area of forest

Table 3.2. The Morisita Index of Similarity between all trees ≥ 1 cm dbh (saplings, poles, and trees) in five plots at Furu. Index values close to 1 indicate high similarity in tree species composition between the sites or plots.

	Plot 1	Plot 2	Plot 3	Plot 4	Plot 5
Plot 1	0	0.60	0.65	0.36	0.08
Plot 2		0	0.38	0.08	0.04
Plot 3			0	0.12	0.09
Plot 4				0	0.84

are required to adequately assess the diversity of vegetation at these sites. Species-area curves demonstrate that sampling 0.5 ha with Whittaker plots and transects is insufficient for documenting species diversity in areas like Papua where tropical forests contain a rich and spatially variable plant assemblage. A minimum of 1 ha (10 Whittaker plots or 4 transect lines) should be sampled during future RAP surveys.

Unfortunately logistical constraints (it is difficult to find large "flat" areas in Papua where plots can rapidly be established) and time constraints may preclude this at many sites. During this RAP survey, Whittaker plots required more time to establish than transect lines but provided a better description of vegetation of the area because they "forced" us to sample seedlings, grasses, and herbaceous plants which are often overlooked when sampling transects. Unfortunately we were unable to identify many of the grasses, seedlings, and saplings in the plots and future RAP surveys would benefit greatly from the inclusion of a botanist with skills in their identification, and/or in the identification of herbaceous plants, ferns, or mosses.

CONSERVATION RECOMMENDATIONS

Threats to biodiversity in the Mamberamo Basin include the nearly complete trans-Papuan highway (from Jayapura to Wamena), which will provide improved access to the region. This may lead to sharp increases in resource exploitation including illegal logging and uncontrolled hunting. Oil palm plantations are currently being developed in nearby regions, and if the Mamberamo dam project proceeds it will flood vast areas of lowland forest. The forests at Furu and Tiri support an exceptional floral diversity that compares favorably with many other forests in the old-world tropics. It is cause for alarm that only a tiny fraction of lowland rain forest in the province occurs in existing protected areas. On the basis of these facts alone forests in the Dabra area warrant some form of protection in the future. Low abundance of individual species in the region dictates that the area to be protected must be relatively large to incorporate viable populations that will ensure long-term survival of many low-density plant species. Fortunately, in the Dabra area the low density of trees normally harvested by commercial logging operations may render these forests economically unprofitable for logging unless operations harvest a wider than normal range of species. Initiatives to protect this region must involve the participation of local communities. Communities in and around Dabra have already shown a significant interest in and support for forest conservation suggesting that long-term conservation initiatives in this forest have a good chance of success.

LITERATURE CITED

Kameubun, K.M. 2000. Keanekaragaman Jenis Pohon Hutan Hujan Tropis dataran rendah di Desa Yemang, Yongsu Dosoyo, Kecamatan Depapre, Kabupaten Jayapura. Unpublished S1 Skripsi. Jayapura; Universitas Cenderawasih.

Krebs, C. 1989. Ecological Methodology. New York: Harper Collins.

Johns, R.J. 1995. Malesia—An Introduction. Curtis's Botanical Magazine 12 (2): 52–62.

Myers, N. 1988. Threatened biotas: 'hotspots in tropical forests.' Environmentalist 8:187–208.

Myers, N., R.A. Mittermeier, C.G. Mittermeier, G.A.B. da Fonseca, and J. Kent. 2000. Biodiversity hotspots for conservation priorities. Nature 403: 853–858.

Paijmans, K. (ed.). 1976. New Guinea Vegetation. Canberra: Australian National University Press.

Stohlgren, T.J. 1995. A modified-Whittaker nested vegetation sampling method. Vegetatio. 117: 113–121.

Whitmore, T.C. 1990. An Introduction to tropical rain forests. Oxford: Oxford University Press.

Chapter 4

Aquatic Insects of the Dabra area, Mamberamo River Basin, Papua, Indonesia

Dan A. Polhemus

CHAPTER SUMMARY

- Aquatic insects were sampled at nine stations in the Dabra area of the Mamberamo River Basin between 5 and 15 September 2000.

- At least 40 species in 23 genera of aquatic Heteroptera were collected, of which at least 17 are new to science.

- Fourteen species in nine genera of Zygoptera, all previously described, and two species in two genera of Gyrinidae, with the number of new species uncertain pending detailed analysis, were collected.

- Aquatic insect diversity in the Dabra area is high, and similar to that seen at corresponding elevational zones in the Wapoga River Basin, which lies immediately to the west of the Mamberamo.

- The aquatic ecosystems of the Dabra area are in excellent condition, but care must be taken with future land use choices, particularly clearing of steep slopes for gardens, to avoid degrading these important resources.

RINGKASAN BAB – SERANGGA AIR DABRA

- Data serangga air diperoleh dari sembilan stasiun di daerah Dabra, daerah aliran Sungai Mamberamo antara 5 dan 15 September 2000.

- Sedikitnya 40 spesies dari 23 genus Heteroptera air berhasil dikumpulkan; dari jumlah tersebut, sedikitnya terdapat 17 spesies baru bagi ilmu pengetahuan.

- Berhasil dikumpulkan empat belas spesies dari 9 genus Zygoptera, yang semuanya telah dideskripsikan, dan 2 spesies dari 2 genus Gyrinidae, dengan kemungkinan sejumlah spesies baru yang memerlukan analisa lebih detil.

- Keanekaragaman serangga air di Dabra tergolong tinggi, dan mirip dengan yang terlihat pada zona elevasi yang sama di daerah aliran Sungai Wapoga, yang terletak di sebelah barat Mamberamo.

- Kondisi ekosistem air tawar di daerah Dabra tergolong luar biasa, tapi perlu dipelihara melalui pemilihan pola pemanfaatan lahan di masa mendatang, khususnya penggundulan lereng yang curam untuk kebun, agar penurunan sumberdaya yang penting ini dapat terhindar.

INTRODUCTION

The current RAP survey of aquatic insects was intended to provide an initial biodiversity profile of selected groups occurring in the Dabra area, in conjunction with ongoing conservation and land use planning initiatives in the central Mamberamo River Basin. As with previous aquatic insect surveys undertaken in the Kikori, Ajkwa, and Wapoga river basins of Papua New Guinea and Papua, the primary groups utilized were aquatic Heteroptera (true bugs), Zygoptera (damselflies), and Gyrinidae (whirlygig beetles). These groups were selected because of their consistency of representation across a wide range of altitudes and habitat types, variation of species assemblages on a local scale between sampling sites, and relatively well investigated taxonomy. The latter factor in particular allowed confidence that identifications could be made to at least the genus level, and that the potential number of undescribed species would be kept to a minimum. The inability to identify target taxa is a factor that frequently limits the utility of studies on tropical insects.

METHODS

Brief but intensive surveys were made at nine sampling stations in two localities near Dabra: the Furu River, a tributary of the Idenburg River lying 3 km SE of Dabra, and the Tiri River, a tributary of the Doorman River, which enters the Idenburg River downstream of Dabra. The sampling stations are described in detail in Appendix 12.

Heteroptera, Zygoptera, and Gyrinidae were collected intensively at each sampling station, while other taxa such as Dytiscidae, Ephemeroptera, and Trichoptera were collected on an opportunistic basis. Collections in the Dabra area were made by visual searching, hand netting, and localized pyrethrin fogging of riparian logs and hygropetric habitats. In addition, limited collections were made from several sites in the Cyclops Mountains where the initial training phase of the RAP was conducted. These sites are described in Appendix 15. Specimens were preserved in 75% ethanol and transported to the Smithsonian Institution for detailed examination and identification. Specimens from these collections will eventually be divided between the Smithsonian Institution and the Indonesian Institute of Sciences (LIPI) in Cibinong, Java.

RESULTS AND DISCUSSION

The Dabra RAP survey documented 56 species of aquatic insects including 40 Heteroptera, 14 Zygoptera, and 2 Gyrinidae. These taxa are listed in Appendix 13 and additional comments on each species are provided in Appendix 14. At least 17 of the species collected represent taxa new to science. Results from the limited Cyclops Mountains collections are presented in Appendix 15.

The aquatic environments at Dabra show many similarities to those sampled during the 1998 Wapoga Basin RAP. Elevations of the Dabra sites (50–100 m asl) are intermediate between sites surveyed at Wapoga: Siewa, at 80 m (3°02'40" S, 136°22'20" E), and Logari River at 275 m (3°00'21" S, 136°33'20" E). This general similarity in elevation and habitat type is reflected in the moderate number of aquatic insects shared among the three areas. The raw species totals for the three groups studied in detail at Dabra also reflect the fact that the Dabra area is topographically and ecologically intermediate between Siewa and Logari; total species diversities for these three sites form a continuum of gradually declining diversity with increasing elevation (Table 4.1). Although lower elevation sites are more diverse, it must be noted that local endemism also increases with increasing elevation along the northern flank of the New Guinea mountains up to at least 1000 m., so that higher sites, although slightly less rich, are equally if not more biologically significant.

A comparison of damselflies (Zygoptera) captured by the 3rd Archbold expedition in the vicinity of Bernhard Camp, and the fauna recorded around Dabra, is of interest because both sites are in roughly the same ecological zone. The Archbold group spent 10 months at Bernhard Camp and collected 35 species of Zygoptera, while the current RAP team spent 11 days in the Dabra area and captured 14 species (Table 4.2). All species taken at Dabra were also collected in the Bernhard Camp area (Table 4.2), indicating that the ecological similarity between the two areas is high. The relatively low diversity of zygopterans at Dabra compared to Bernhard Camp suggests that rapid surveys during a single season may not be adequate to characterize the diversity of Zygoptera at sites subject to strong seasonal variations in rainfall.

Another notable aspect of the aquatic entomofauna around Dabra was the presence of large palingeniid mayflies in the genus *Plethogenesia*, which underwent a spectacular hatch on the morning of 6 September 2000. Thousands of these large, yellowish mayflies emerged onto the river surface where they flew at speeds equal to or greater than that of a motorized dugout. Immature stages, which burrow into soft sediments of river banks and river channels, were obtained from local boatmen who were using them for fish bait. The presence of these burrowing mayflies is indicative of a river

Table 4.1. Comparison of species totals for selected aquatic insect groups at Dabra and ecologically similar sites in the Wapoga River Basin (from Polhemus, 2000).

Taxon	Siewa	Dabra	Logari
Heteroptera	47	42	27
Zygoptera	16	14	10
Gyrinidae	3	2	2

system without major sources of chemical pollution, since they are quickly lost in areas subjected to industrial development (Polhemus, 1994).

Overall, the stream catchments of the Dabra area are in excellent condition, despite a certain amount of selective

logging that has taken place along the Furu River. High species totals for Heteroptera and Zygoptera are notable, considering that at Siewa and Logari human presence was minimal to non-existent, while at Dabra there is a moderate local population density. Given numerous steep slopes in the

Table 4.2. Comparison of Zygoptera diversity at present RAP sites versus Archbold Expedition collections at Bernhard Camp (Archbold et al., 1942).

Taxon	Bernhard Camp	Furu	Tiri
Neurobasis australis	X	X	X
Neurobasis ianthipennis	X	-	-
Rhinocypha tincta amanda	X	X	X
Lestes lygisticercus	X	-	-
Podopteryx casuarina	X	-	-
Argiolestes amphistylus	X	-	-
Argiolestes aulicus	X	-	-
Argiolestes lamprostomus	X	-	-
Argiolestes simplex	X	-	-
Argiolestes subornatus	X	-	-
Drepanosticta clavata	X	-	-
Drepanosticta lepyricollis	X	-	-
Nososticta beatrix	X	-	X
Nososticta erythrura	X	-	X
Nososticta fonticola	X	-	-
Nososticta irene	X	X	-
Nososticta melanoxantha	X	X	X
Nososticta nigrofasciata	X	X	X
Nososticta plagioxantha	X	-	-
Selysioneura phasma	X	X	-
Tanymecosticta fissicollis	X	-	-
Idiocnemis obliterata	X	X	-
Palaiargia ceyx	X	-	X
Papuagrion occipitale	X	X	-
Teinobasis argiocnemis	X	-	-
Teinobasis dominula	X	-	-
Teinobasis olthofi	X	-	-
Teinobasis serena	X	-	-
Teinobasis rufithorax	X	-	-
Pseudagrion civicum	X	X	X
Pseudagrion farinicolle	X	X	-
Pseudagrion pelecotomum	X	-	-
Archibasis crucigera	X	-	-
Argiocnemis ensifera	X	X	-
Ischnura stueberi	X	-	-
Total	**35**	**11**	**8**

Dabra area, extensive forest clearance beyond that currently undertaken for gardens along the valley floors will have rapid and detrimental effects on local stream ecosystems, degrading water quality and affecting the diversity of animal communities. Such effects may already be seen around gardens to the northwest of Dabra, where streambeds have a scoured appearance due to loss of forest cover on surrounding slopes. It will require wise land use choices to maintain the streams of this area in their current state of good health.

LITERATURE CITED

Archbold, R., A.L. Rand, and L.J. Brass. 1942. Results of the Archbold Expeditions. No. 41. Summary of the 1938–1939 New Guinea Expedition. Bull. Am. Mus. Nat. Hist. 79: 197–288.

Polhemus, D.A. 2000. Aquatic insects of the Wapoga River area, Irian Jaya, Indonesia. *In:* Mack, A. and L.E. Alonso (eds.). A biological assessment of the Wapoga River area of northwestern Irian Jaya, Indonesia. RAP Bulletin of Biological Assessment 14. Washington, DC: Conservation International. Pp. 39–42.

Polhemus, D.A. 1994. Conservation of aquatic insects: worldwide crisis or localized threats? Am. Zool. 33: 588–598.

Chapter 5

Butterflies of the Yongsu area, Papua, Indonesia

*Edy Rosariyanto, Henk van Mastrigt,
Henry Silka Innah, and Hugo Yoteni*

CHAPTER SUMMARY

- 69 species of butterflies in four families were positively identified in the Yongsu area.

- One species, *Elymnias paradoxa* (Nymphalidae), was previously known only from eastern Papua New Guinea.

- Nocturnal butterflies were more abundant at 150 m asl than at 70 m asl.

RINGKASAN BAB – KUPU-KUPU YONGSU

- 69 spesies kupu-kupu dari empat famili berhasil diidentifikasi di daerah Yongsu.

- Ditemukan satu spesies, *Elymnias paradoxa* (Nymphalidae), yang sebelumnya hanya diketahui berada di sebelah timur Papua New Guinea.

- Kupu-kupu malam lebih melimpah pada ketinggian 150 m dpl daripada 70 m dpl.

INTRODUCTION

The butterfly fauna of Papua has been less intensively studied than that of neighboring Papua New Guinea, with about 645 species (excluding Hesperiidae) currently recognised from the province (D'Abrera, 1971, 1977, 1990; Parsons, 1999) compared to 785 species in PNG (Parsons, 1999). A number of recent studies in the province have focused on the predominantly montane genus *Delias* (e.g. van Mastrigt, 1990, 1996, 2001), but there have been few recent surveys of lowland faunas. The aim of this study was to document the butterfly fauna of coastal lowland forests on the northern side of the Cyclops Mountains.

METHODS

We sampled day flying butterflies for three days in closed forest and three days in village gardens and forest edge between 23 and 29 August 2000. Butterflies were collected with long-handled nets between 9:00 a.m. and 1:00 p.m., anaesthetised with an ammonia solution, and stored in paper envelopes. Nocturnal butterflies were sampled on six nights from two altitudes (70 m asl and 150 m asl). Their abundance was documented by counting the number of individuals attracted to fluorescent and mercury vapour lights in primary forest at 8:30 p.m., 9:30 p.m., and 10:30 p.m.

RESULTS AND DISCUSSION

Sixty-nine species of butterflies from four families were recorded during the survey: Papilionidae (7 species), Pieridae (7 species), Lycaenidae (17 species) and Nymphalidae (38 species) (Appendix 16). The family Hesperiidae is not considered in this chapter. Diversity was higher in the garden habitat/forest edge than in the forest (56 vs 38 species), possibly as a consequence of low light levels in the closed forest habitat. In addition gardens were sampled on three clear sunny days but one of the forest sampling days was cool and overcast. This may have significantly reduced butterfly activity for one third of the sampling period in this habitat.

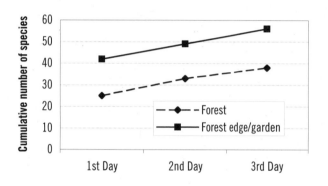

Figure 5.1. Species accumulation curves for day-flying butterflies in forest and in Yongsu village gardens and forest edge.

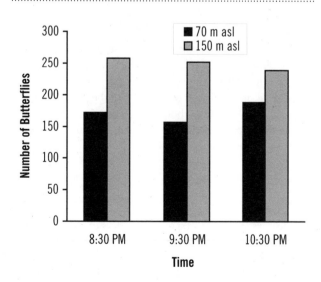

Figure 5.2. Effect of altitude on nocturnal butterfly abundance at Yongsu Camp.

Two species of conservation significance were recorded. The first is *Ornithoptera priamus* (Papilionidae). This large and spectacular butterfly is popular in the insect trade and is protected by Indonesian Government regulations. It is widespread in the New Guinea region and was common at the Yongsu site. There did not appear to be any threats to the Yongsu population. The second significant species is *Elymnias paradoxa* (Nymphalidae). The Yongsu record represents a significant westerly range extension for this species, which was previously known only from eastern PNG (Parsons, 1999).

Documentation of 69 species in the small area accessed during this short survey demonstrates that the butterfly fauna at Yongsu is exceptionally rich. Total diversity at the site represents nearly half the 144 species reported from the entire Cyclops Nature Reserve (van Mastrigt, 1984). Furthermore, species accumulation curves (Figure 5.1) indicate that additional species are likely to occur in both the forest and in more open, sunny areas. More individuals of nocturnal butterflies were found at 150 m than at 70 m asl (Figure 5.2). This may be a real altitudinal effect, or it may reflect increasing distance from the coast. In either case additional surveys at a range of altitudes would provide useful information about patterns of butterfly diversity across altitudinal gradients in the Cyclops Mountains. Along with information about the proportion of rare or locally restricted species at different altitudes, these data should be used for identifying priority areas for butterfly conservation within the Cyclops Mountains area.

LITERATURE CITED

D'Abrera, B. 1971. Butterflies of the Australian Region, 1st edition. Melbourne: Lansdowne Press.

D'Abrera, B. 1977. Butterflies of the Australian Region, 2nd edition. Melbourne: Lansdowne Press.

D'Abrera, B. 1990. Butterflies of the Australian Region, 3rd edition., Melbourne: Hill House Publishers.

Parsons, M. 1999. The Butterflies of Papua New Guinea: Their Systematics and Biology. London: Academic Press.

van Mastrigt, H.J.G. 1984. Butterflies of the Cyclops Mountains area. *In* Ratcliffe, J.B. (compiled) Cagar Alam Pegunungan Cyclops, Irian Jaya, Indonesia. Management Plan 1985–1989. Jayapura: World Wildlife Fund. Pp. 70–74.

van Mastrigt, H.J.G. 1990. New (sub)species of *Delias* from the central mountain range of Irian Jaya (Lepidoptera, Pieridae). Tijd. Entom. 133: 197–204

van Mastrigt, H.J.G. 1996. New species and subspecies of *Delias* Hübner 1819 from the central mountain range of Irian Jaya, Indonesia (Lepidoptera, Pieridae). Neue Entom. Nach. 38: 21–55.

van Mastrigt, H.J.G. 2001. Study on *Delias* from lower montane forest in Irian Jaya, Indonesia (Lepidoptera, Pieridae). Futao. 37: 2–13.

Chapter 6

Butterflies and Moths of the Dabra area, Mamberamo River Basin, Papua, Indonesia

Henk van Mastrigt and Edy M. Rosariyanto

CHAPTER SUMMARY

- 129 species of butterflies were collected at 2 camps.

- The butterfly fauna was dominated by species typical of lowland coastal areas around Jayapura and/or Nabire.

- More than 480 species of moths in over 112 genera were recorded. Moths of the family *Pyralidae* were particularly diverse with over 145 species, many of which remain unidentified. Large moths of the families *Sphingidae* (only 5 species) and *Saturniidae* (none) were very poorly represented.

RINGKASAN BAB – KUPU-KUPU DAN NGENGAT DABRA

- 129 spesies kupu-kupu berhasil diperoleh dari 2 lokasi penelitian.

- Kupu-kupu didominasi oleh spesies yang tipikal untuk kawasan pesisir dataran rendah di sekitar Jayapura dan/atau Nabire.

- Lebih dari 480 spesies Ngengat yang berasal lebih dari 112 genus berhasil dicatat. Ngengat dari famili Pyralidae cukup beragam dengan lebih dari 145 spesies, kebanyakan masih belum teridentifikasi. Ngengat besar dari famili Sphingidae (hanya 5 spesies) dan Saturniidae (tidak ada) kurang terwakili.

INTRODUCTION

Approximately 803 species of butterflies are known from New Guinea (D'Abrera, 1971, 1977, 1990; Parsons, 1999). There are substantial differences of opinion among lepidopterists about the status of many taxa, and some "species" previously considered to have wide distributions have recently been split into several taxa with more restricted distributions. These developments have led to the recognition of an increasing number of taxa and of more locally endemic species.

Different opinions also occur about the number of butterfly families in New Guinea. D'Abrera (1971, 1977, 1990) recognizes nine families, but Parsons (1999) recognizes only four (Papilionidae, Pieridae, Nymphalidae, and Lycaenidae) divided into fifteen subfamilies. Parsons (1999) also treats the family Hesperiidae as "true" butterflies but D'Abrera does not, and we do not consider this group in this chapter. These taxonomic differences aside, however, it is clear that the butterfly fauna of Papua is very diverse and includes a large number of very spectacular species.

The aim of this survey was to document the butterfly fauna of the Dabra region in the Mamberamo River Basin, to compare this fauna with that known from other lowland sites, and to identify any conservation issues for the Mamberamo fauna.

METHODS

During two weeks in the Mamberamo River Basin near the village of Dabra, butterflies were collected in the vicinity of two sites:

Furu camp (3°17'04"S, 138°38'9.5"E, 2–7 September, 2000): Furu camp was at the northern base of Papua's central mountains, and streams in the area ranged from low to very high gradient. South of camp, the forest—including the borders of the rivers—was in good condition (primary forest). North of camp, however, the forest was extensively disturbed and included open areas and gardens. These differences were reflected in different assemblages of butterflies; larger numbers of common and widely distributed butterflies were found in the disturbed areas. Fewer species were found in the primary forest areas, but they included a number of less-common species.

Tiri camp (3°17'30"S, 138°34'54.2"E, 9–14 September, 2000): Tiri camp was in flatter forest. The soil was moister, and the forest was in good condition (primary forest). Large gardens occurred along the river about a 40-minute walk north of the camp, and further north the forest became more open. However, we spent little time at these two disturbed localities. Weather conditions were generally similar at both camps and are unlikely to be responsible for any differences in the fauna observed at the two sites.

Butterflies were surveyed by three people from 1 to 14 September, 2000, using long-handled nets. We generally collected between 9 a.m. and 2 p.m., although collecting effort varied according to weather conditions. Butterfly activity was low on cloudy mornings, so collecting effort was reduced during those conditions. Each day we recorded the total number of species collected and observed. We also made observations on the behavior of butterflies in the wild. A small series of voucher specimens was retained and is deposited in the collection of Henk van Mastrigt in Jayapura. Most specimens, including all protected species, were released unharmed after identification.

Moths were surveyed every night between 6 and 10 pm with one (Tiri) or two (Furu) 160 Watt white lamps placed in front of large white screens. At Furu one lamp was placed about 10 meters from the river on an elevated bank, and the second one was placed next to and on the same level as the river. At Tiri the lamp was placed next to and on the same level as the river. Specimens were anaesthetized in killing jars containing potassium cyanide. Larger moths were killed by an injection of 5% ammonia. After moths were set and dried, photographs were taken and these were eventually sent to scientists specializing in one or more genera or groups.

Their comments form the basis for many of the moth identifications reported here.

Butterflies were identified using the monographs of D'Abrera (1971, 1977, 1990) and Parsons (1999). Unfortunately no single reference is available to identify the moths of Papua, and most of the primary literature is scattered and difficult to obtain. However, a number of monographs that focus on South-east Asian moths were extremely useful for preliminary identifications of Papuan taxa (Barlow, 1989; D'Abrera, 1995, 1996; Holloway, 1983, 1985, 1986, 1987, 1988, 1996, 1997, 1998, 1999; Holloway et al., 2001; Robinson et al., 1994).

RESULTS

Butterflies

One hundred and twenty-nine species in 68 genera were recorded (Table 6.1; Appendix 17). Most are common and widespread species typical of the lowland forests around Nabire and/or Jayapura. The results of this survey are compared with the total known lowland Papuan fauna (358 species in 111 genera and 15 subfamilies - 4 families) in Table 6.1. Given the paucity of comprehensive lepidopteran surveys at other lowland sites in Papua, it is difficult to assess whether the diversity of butterflies at the Mamberamo sites is low or high. The totals for both Furu (109) and Tiri (89) are higher than the total diversity (68) documented during a mini-survey at the northern base of the Cyclops Mountains from 20 to 30 August, 2000 (Rosarianto et al., this volume).

The low percentage of the total Papuan lowland butterfly fauna encountered around the Mamberamo camps (26.8%) suggests that one short, dry-season survey is inadequate to document the fauna at these sites. Species accumulation curves (Figure 6.1) suggest that a longer survey would have encountered many more species, and a wet-season survey would undoubtedly increase the known diversity of the region substantially.

Moths

Over 480 species of moths were collected in the Dabra area (Appendix 18). Given an expected butterflies: moths ratio of about 1:7 (Smart, 1977) the diversity of moths recorded from this area appears to be quite low, possibly due to the lack of elevated, exposed sites that are used for attracting moths to lights. However, previous surveys in the lowlands and mountains of Papua by the senior author have documented much higher abundances (5 to 10 times higher) of specimens but much lower diversity (van Mastrigt, unpublished), suggesting that the Mamberamo fauna is quite diverse compared to other sites in the Province. Several groups including the families Sphingidae and Saturniidae were poorly represented in the collection: only eight specimens of Sphingidae (five different taxa) and no Saturniidae at all. In contrast a large number of specimens and species of the family Pyralidae were found, especially in the subfamily Pyraustinae (106 species). The subfamily Acentropinae, a group associated with wet environments, was present in overwhelming numbers but was represented by only three of the nineteen species in this group.

DISCUSSION AND CONCLUSIONS

Although more surveys are required to obtain a comprehensive overview of lepidoptera in the Mamberamo area, the butterfly assemblage shows similarities with other lowland faunas on the northern side of Papua. It contains many components of assemblages occurring at Nabire (to the southwest) and Jayapura (to the northeast), but there appears to be somewhat more similarity with the Jayapura fauna. For example, *Parthenos aspila* and *Elymnias thryallis* are common to the Mamberamo area and the vicinity of Jayapura, while two different species of these genera (*Parthenos tigrana* and *Elymnias agondas*) occur in the Nabire area. Some species that are common around Jayapura and Nabire, and that we expected to find in the Mamberamo area, were not found during this survey. These included *Papilio albinus, Papilio*

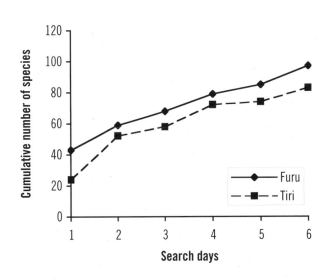

Figure 6.1. Species accumulation curves for butterflies at Furu and Tiri Camps in the Mamberamo basin, Papua, Indonesia

Table 6.1. Butterflies (excluding Hesperiidae) recorded during the present survey compared with total number of lowland species known from mainland Papua.

Family	Subfamily	Mainland Papua (lowland only)		Mamberamo survey		% of total known species	
		Genera	Species	Genera	Species	Genera	Species
Papilionidae	Papilioninae	5	22	5	12	100	54.6
subtotal		5	22	5	12	100	54.5
Pieridae	Coliadinae	3	10	2	4	66.7	40.0
	Pierinae	6	16	4	7	66.7	43.7
subtotal		9	26	6	11	(ave) 66.7	(ave) 42.3
Lycaenidae	Riodininae	2	8	1	1	50.0	12.5
	Curetinae	1	1	1	1	100	100
	Lycaeninae	48	172	22	48	45.8	27.9
subtotal		51	181	24	50	(ave) 47.1	(ave) 27.6
Nymphalidae	Libytheinae	1	2	1	1	100	50.0
	Ithomiinae	1	4	1	1	100	25.0
	Danainae	5	20	4	8	80.0	40.0
	Morphinae	3	14	2	5	66.7	35.7
	Satyrinae	8	39	3	12	37.5	30.8
	Charaxinae	3	3	3	3	100	100
	Apaturinae	4	5	2	3	75.0	60.0
	Nymphalinae	12	31	11	16	91.7	51.6
	Heliconiinae	9	11	7	7	77.8	63.6
subtotal		46	129	34	56	(ave) 73.9	(ave) 43.4
Total		111	358	69	129	(ave) 62.2	(ave) 36.0

fuscus, Graphium macfarlanei, and the introduced *Papilio demoleus.*

Moths are considered to be excellent biodiversity indicators due to their abundance, diversity, and different assemblage structure in different types and qualities of habitats. However, to be useful indicators a minimal level of identification is necessary, and this requires training in identification of moth groups. Providing such training for several Papuan conservation biologists would greatly enhance future biodiversity assessments in Papua.

In summary, the overall diversity and abundance of the lepidopteran fauna indicates that the forests in the Mamberamo area are still in good condition. Further studies are required to determine the total diversity of the fauna and to assess the status of rare and protected species.

LITERATURE CITED

Ackery, P.R. and R.I. Vane-Wright. 1984. Milkweed Butterflies. London: British Museum (Natural History).

Barlow, H.S. 1982. An Introduction to the Moths of South East Asia. The Malaysian Nature Society. Kuala Lumpur: Art Printing Works.

D'Abrera, B. 1971. Butterflies of the Australian Region, 1st edition. Melbourne: Lansdowne Press.

D'Abrera, B. 1977. Butterflies of the Australian Region, 2nd edition. Melbourne: Lansdowne Press.

D'Abrera, B. 1990. Butterflies of the Australian Region, 3rd edition. Melbourne: Hill House Publishers.

D'Abrera, B. 1995. Saturniidae Mundi, Part I. Melbourne: Hill House Publishers.

D'Abrera, B. 1998. Saturniidae Mundi, Part III. Keltern: Goecke and Evers.

Holloway, J.D. 1983. The moths of Borneo, part 4. Malay. Nat. J. 37: 1–107.

Holloway, J.D. 1985. The moths of Borneo, part 14. Malay. Nat. J. 38: 157–317.

Holloway, J.D. 1986. The moths of Borneo, part 1. Malay. Nat. J. 40: 1–165.

Holloway, J.D. 1987. The moths of Borneo, part 3. Kuala Lumpur: Southdene Sdn. Bhd.

Holloway, J.D. 1988. The moths of Borneo, part 6. Kuala Lumpur: Southdene Sdn. Bhd.

Holloway, J.D. 1996. The moths of Borneo, part 9. Malay. Nat. J. 49: 147–326.

Holloway, J.D. 1997. The moths of Borneo, part 10. Malay. Nat. J. 51: 1–242.

Holloway, J.D. 1998. The moths of Borneo, part 8. Malay. Nat. J. 52: 1–155

Holloway, J.D. 1999. The moths of Borneo, part 5. Malay. Nat. J. 53: 1–188.

Holloway, J.D., G. Kibby, and D. Peggie. 2001. The families of Malesian moths and butterflies. Fauna Malesian Handbook 3. Leiden: Brill.

Otani, Takuya and Yasuhiko Kimura. 1998. Birdwing Butterflies. Tokyo: Toshitsugu Endo.

Parsons, M. 1999. The Butterflies of Papua New Guinea: Their Systematics and Biology. London: Academic Press.

Robinson, G. S., K. R. Tuck, and M. Shaffer. 1994. A Field Guide to the Smaller Moths of South-East Asia. London: The Natural History Museum.

Smart, P. 1977. The illustrated Encyclopedia of the Butterfly World in Colour. London: Hamlyn.

Chapter 7

Fishes of the Yongsu and Dabra areas, Papua, Indonesia

Gerald R. Allen, Henni Ohee, Paulus Boli, Roni Bawole, and Maklon Warpur

CHAPTER SUMMARY

- The Yongsu freshwater fish fauna consists of 33 species in 25 genera and 15 families.

- Yongsu fishes are adapted to relatively steep-gradient streams rising a short distance from the sea. The fauna is similar to other assemblages inhabiting mountainous coastlines on the north coast of New Guinea.

- The Yongsu fauna is dominated by gobioid fishes (Gobiidae and Eleotridae), which account for nearly half of the total fishes. The cling-goby subfamily Sicydiinae is particularly well represented with seven species.

- The Yongsu area and adjacent Cyclops coast may provide the best example of steep-gradient coastal stream habitat on the entire mainland of Papua. It is also the home of two apparently endemic gobies in the genus *Lentipes*, including a new species discovered during this training survey, thus providing justification for conservation initiatives.

- The Mamberamo fauna consists of 35 known species of which 23 species (in 18 genera and 11 families) were recorded during the present survey.

- The Mamberamo fauna is most closely related to that of the Sepik and Ramu rivers of Papua New Guinea. About 70 percent of Mamberamo fishes also occur in these two rivers. There is also a significant endemic element with six species restricted to the Mamberamo Basin.

- The Mamberamo has both the highest percentage of endemic fishes and the highest percentage of introduced fishes of any major river system in New Guinea.

- Additional ichthological surveys in the Mamberamo Basin are urgently required.

RINGKASAN BAB – IKAN-IKAN AIR TAWAR YONGSU DAN DABRA

- Di Yongsu ditemukan 33 spesies ikan air tawar, yang mewakili 25 genus dan 15 famili.

- Ikan-ikan di Yongsu telah teradaptasi dengan kondisi sungai yang relatif curam yang jaraknya dekat dari laut. Fauna ikannya mirip dengan kelompok-kelompok lain yang menghuni pesisir yang berbatasan dengan pegunungan di pesisir utara New Guinea.

- Fauna ikan di Yongsu didominasi oleh ikan-ikan gobi (Gobiidae dan Eleotridae), yang mencakup hampir setengah dari jumlah total ikan. Ikan *cling-goby* sub-famili Sicydiinae terwakili dengan baik oleh tujuh spesies.

- Daerah Yongsu dan sekitar pesisir Cyclops merupakan contoh terbaik untuk habitat sungai pesisir yang curam di seluruh daratan Papua. Daerah ini juga merupakan habitat bagi dua spesies gobi endemik dari genus *Lentipes,* salah satunya merupakan spesies baru yang ditemukan saat pelatihan, sehingga dapat menjadi justifikasi untuk inisiatif konservasi.

- Jumlah fauna ikan yang telah diketahui berada di Mamberamo adalah 35 spesies, namun yang berhasil dicatat dalam survei ini adalah 23 spesies mewakili 18 genus dan 11 famili.

- Fauna ikan Mamberamo sangat mirip dengan yang ada di Sungai Sepik dan Ramu di Papua New Guinea. Sekitar 70% ikan-ikan di Mamberamo juga terdapat pada kedua sungai tersebut. Tingkat endemisitasnya juga cukup berarti dengan adanya enam spesies ikan yang hanya ditemukan di daerah aliran Sungai Mamberamo.

- Persentase ikan endemik dan ikan introduksi di Mamberamo adalah tertinggi dibandingkan sungai-sungai utama di New Guinea.

- Survei-survei lanjutan mengenai Ichtyologi/ikan-ikan di daerah aliran Sungai Mamberamo sangat diperlukan.

INTRODUCTION

New Guinea freshwater fishes were most recently summarized by Allen (1991), who documented 330 species from the mainland and nearby islands. Additional discoveries during the past decade have increased this total to about 385 species. The fauna consists mainly of marine-derived groups such as ariid and plotosid catfishes, atherinoids, terapontid grunters, and gobioids. It is further characterized by a high level of endemism (about 60 percent). About 35 species are shared with northern Australia, reflecting the historical land connection between these areas.

Much of New Guinea, including vast areas of Papua, remains unsurveyed and basic information on faunal distribution and diversity is urgently required for conservation planning and management. The only previous collections from the Cyclops Nature Reserve (Adjacent to Yongsu), were made in August 1995 by G. Allen, S. Renyaan, and D. Price at the Omamerwai and Krimpon Rivers, about 12 km west of the Yongsu area. The collections contained 17 species. The first scientific fish collections from the Mamberamo were made by van Heurn in 1920–21. These collections were mainly described by Weber and De Beaufort (1911–1962), although Collette (1982) later described a garfish from this same expedition. A few fishes, including the first specimens of *Gobius tigrellus,* were collected by the American Archbold Expedition in the late 1930s. Several locations within the Mamberamo catchment, including Danau Biru, Obogwi, Faui, Kordesi, Dabra, Nevere, and Senggi were surveyed by G. Allen between 1982 and 1991. Museum specimens from the various expeditions are mainly deposited at the American Museum of Natural History (New York), Museum Zoologicum Bogoriense (Bogor), Western Australian Museum (Perth), and the Zoological Museum (Amsterdam).

This report presents the results of fish surveys undertaken during a Conservation International training course and biodiversity Rapid Assessment Program (RAP) in the Yongsu area near Jayapura and in the vicinity of Dabra in the Mamberamo River Basin.

METHODS

Fishes were collected in the Yongsu area between 23 and 29 November, and at Dabra between 2 and 8 September 2000. A variety of methods were used to sample the fauna. Species from the Yongsu and Dabra areas are reasonably well represented in museum collections, so emphasis was placed on visual surveys rather than collecting. This method was particularly effective around Yongsu where streams were generally clear, facilitating underwater observations with mask and snorkel. Selective collecting was undertaken with small hand nets.

The ichthyocide rotenone was used at a few sites. This chemical is derived from the derris plant, and is ideal for collecting fishes in small creeks or sections of larger streams where flow is minimal. Approximately 0.5–1.0 kg of rotenone powder was mixed with several liters of water and then dispersed into the stream over a period of 10–15 minutes. After several minutes stunned fishes began to gasp at the surface and were easily netted. Rotenone stations were generally in small creeks near their junction with larger streams. In this way a minimal but representative sample of fishes was obtained and the rotenone was diluted rapidly in the flow of the larger stream.

A 15 m fine-meshed seine with a width or "drop" of about 2.0 m was used occasionally in larger creeks. It was weighted with lead on the bottom edge and had a number of floats on the upper edge. Sampling with the seine involved one person in a more or less stationary position on the bank and a second person who waded into the stream and then returned along a roughly U-shaped path. Both ends of the net were then quickly hauled ashore. Descriptions of all sampling sites are presented in Appendix 19. The field team who assisted with fish sampling at one or both sites was: Hendrite Ohee, Roni Bawole, Paulus Boli, and Maklon Warpur.

Voucher specimens were fixed in a 10% formalin solution and later transferred to 75% ethanol for permanent storage in museum collections. Most were deposited at Museum Zoologicum Bogoriense, Bogor, Indonesia (MZB), but a small representative collection and selected specimens that were fixed in ethyl alcohol for future DNA analysis are registered at the Western Australian Museum, Perth (WAM).

Colour photographs of most species were taken in the field. Small fishes, such as rainbowfishes, gobies, and gudgeons, were photographed alive in a small landscaped aquarium. Larger species, such as ariid catfishes, were photographed out of water while fresh. Photography equipment consisted of a Nikon F-801 camera, Nikon 60 mm macro lens, and Nikon SB-26 strobe. The resultant slides were scanned using a Nikon LS-2000 scanner, and compact discs containing a variety of images have been deposited at CI offices in Washington, DC, Jakarta, and Jayapura.

RESULTS AND DISCUSSION

Freshwater fishes of the Yongsu area

Thirty-three species in 25 genera and 15 families were recorded from streams around Yongsu (Appendices 20, 21). The fauna is dominated by gobioid fishes (families Gobiidae, Rhyacichthyidae, and Eleotridae), which account for nearly half of the fishes in the Yongsu area. The gobiid subfamily Sicydiinae was especially prominent in clear, rocky streams, the dominant aquatic habitat in the area. These fishes, commonly known as cling gobies, possess a peculiar "sucking disk," a modification formed by the fused pelvic fins that is used for clinging to rocks in fast-flowing streams. Most of the non-gobioid fishes were basically marine forms restricted to a narrow zone extending no more than about 400 m from the edge of the sea. They were found near the mouths of rocky streams or in a single brackish pool just above the high tide mark.

The Yongsu fauna is similar to that of other mountainous coastal areas of northern New Guinea, especially those on offshore islands such as Yapen and the Raja Ampat group. The Cyclops coast, which is typified by the Yongsu area, harbors a freshwater fish community adapted to relatively steep-gradient streams rising a short distance from the sea. The pelagic larval stage possessed by most coastal-stream species provides a high degree of faunal continuity throughout the region stretching from Waigeo Island to the Solomon Islands.

The only other freshwater fish collections from Cyclops coastal streams were made by G. Allen, S. Renyaan, and D. Price in 1995. They collected five species that were not recorded during the present survey: *Apogon amboinensis* (Apogonidae), *Plectorhinchus gibbosus* (Haemulidae), *Pomacentrus taeniometopon* (Pomacentridae), *Lentipes dimetrodon* (Gobiidae), and *Sicyopterus zosterophorum* (Gobiidae).

In addition to the freshwater survey, the author spent approximately six hours conducting a visual survey of shallow (to 5 m depth) coral reef fishes, using a mask and snorkel. A summary list of species is presented in Appendix 22.

FAUNAL COMPOSITION OF THE DABRA AREA

Twenty-three species in 18 genera and 11 families were recorded during this survey (Appendices 23 and 24), representing 65 percent of the 35 species (including introduced forms) currently known from the Mamberamo drainage. Lack of access to riverine environments prevented a comprehensive inventory, and a number of additional species are likely to occur in these habitats around Dabra. Local informants and a previous collection obtained from Dabra by a Summer Institute of Linguistics (SIL) pilot indicate that at least seven species not recorded during our survey also occur in the area: *Neosilurus idenburgi* (Plotosidae), *Zenarchopterus kampeni* (Hemiramphidae), *Chilatherina crassispinosa* (Melanotaeniidae), *Parambassis confinis* (Ambassidae), mugilid sp. (Mugilidae), *Mogurnda aurofodinae* (Eleotridae), and *Channa striata* (Channidae).

The Mamberamo fauna is most closely related to that of the Sepik and Ramu rivers of Papua New Guinea (PNG; Allen and Coates, 1990; Allen et al., 1992). These three rivers, along with the Markham River of PNG, combine to form the Intermontane Trough or Sepik-Ramu Depression, one of the major structural elements of New Guinea (Löffler, 1977). It extends from Huon Gulf in PNG to Cenderawasih Bay in Papua.

The Trough has been a zone of relative subsidence since the late Tertiary, and its margins are characterized by steep fault scarps. The valley floor is filled with terrestrial clastic sediments forming extensive alluvial plains and fans. The relief between the major river systems is relatively low and judging from the number of shared species, has not formed a significant barrier to fish dispersal (Allen and Coates, 1990; Allen et al., 1992).

The freshwater fauna of the Intermontane Trough is summarized in Table 7.1. Most Mamberamo native species (20 of 28 or 71.4 percent) are also reported from the Sepik and Ramu systems. Exceptions include *Zenarchopterus alleni* (Hemiramphidae), *Chilatherina bleheri, Melanotaenia maylandi, M. praecox* (Melanotaeniidae), *Parambassis altipinnis* (Ambassidae), *Glossamia beauforti* (Apogonidae), and *Eugnathogobius tigrellus* (Gobiidae). Most of these are Mamberamo endemics with the exception of *M. praecox* (also Wapoga system) and *G. beauforti* (also Lake Sentani).

With the exception of some locally endemic taxa, most species currently found in the Ramu-Sepik catchment but not recorded from the Mamberamo can be expected to occur there. Most of these are species with wide distributions that frequently inhabit the lower parts of rivers and adjacent estuaries, habitats that have yet to be adequately sampled in the Mamberamo Basin. The Mamberamo fauna is therefore unlikely to be as impoverished as indicated in Appendix 23. Allowing for an additional 20–25 lower river-estuarine species that are likely to occur in the Mamberamo, the fish diversity of this drainage is probably similar to that of the Sepik-Ramu. The northern rivers of New Guinea are geologically much younger than southern rivers, a fact reflected in their less diversified fauna (Allen, 1991). However, if the entire Intermontane Trough is considered as a single faunal unit, the species total (75, excluding introduced forms) is comparable with that of southern systems (Table 7.1).

Perhaps the most notable record obtained during the present survey was a small (maximum size ~ 3 cm) goby *Eugnathogobius tigrellus*, which was previously known from only 10 poorly preserved specimens collected from the Mamberamo by the Archbold Expedition in 1939. Ten additional specimens were collected from small creeks near Furu and Tiri Camps during the present survey, and the first photographs of this species in life were obtained. The additional specimens will help to clarify this species' generic affinities, which are unclear. It was originally placed in *Gobius*, which has long served as a "catch-all" group for a variety of dissimilar species. However studies by gobiid specialist Helen Larson (Northern Territory Museum, Australia) indicate that

although the species is closely related to *Eugnathogobius*, it may represent an undescribed genus.

Zoogeographic affinities of the Yongsu and Mamberamo fish faunas

The zoogeographic affinities of Yongsu and Mamberamo fishes are summarized in Tables 7.2 and 7.3 respectively. These data illustrate the significant faunal differences between the two areas. The majority (about 85%) of Yongsu fishes are widespread forms whose distributions encompass either the tropical Indo-Pacific, Western Pacific, or Indo-Australian Archipelago. In contrast, the Mamberamo fauna is dominated by Intermontane Trough and northern New Guinea endemics, which comprise about 70 percent of the fauna. Six (17.1%) of the Intermontane Trough species are restricted to the Mamberamo Basin.

Table 7.1. Comparison of fish faunas of various river systems in New Guinea.

River system	Total species (excluding introductions)	Endemic species	Percent endemics
Fly, PNG	103	5	4.8
Kikori , PNG	100	14	14.0
Aikwa/Iwaka, Irian Jaya	75	2	5.4
Lorentz, Irian Jaya	60	2	3.3
Purari , PNG	57	6	10.5
Sepik, PNG	53	3	5.6
Ramu, PNG	50	2	4.0
Wapoga, Irian Jaya	46	3	6.5
Digul, Irian Jaya	40	0	---
Mamberamo, Irian Jaya	**28**	**6**	**17.1**
Gogol, PNG	25	0	---
Lakekamu, PNG *	22	1	4.5

Data taken from unpublished studies by the author and those of Roberts (1978), Weber (1913), Haines (1979), Allen and Coates (1990), Allen et al. (1992), Allen and Boeseman (1982), and Parenti and Allen (1991).

* indicates incompletely surveyed

Table 7.2. Zoogeographic affinities of Yongsu fishes.

Zoogeographic affinities	Number of species (% of fauna)
Indo-West Pacific	14 (42.4)
Western Pacific	6 (18.2)
Indo-Australian Archipelago	8 (24.3)
Northern New Guinea	4 (12.1)
Cyclops Coast, Irian Jaya	1 (3.0)

Table 7.3. Zoogeographic affinities of Mamberamo fishes.

Zoogeographic affinities	Number of species (% of fauna)
Intermontane Trough*	13 (37.1)*
Northern New Guinea	12 (34.3)
Introduced (Africa & SE Asia)	6 (17.1)
Indo-West Pacific	3 (8.6)
New Guinea and northern Australia	1 (2.9)

* includes 6 (17.1%) Mamberamo endemics

CONCLUSIONS AND RECOMMENDATIONS

The Yongsu and Dabra areas provide good examples of two very different types of freshwater fish communities. Coastal stream species with marine egg and pelagic larval stages dominate the Yongsu fauna, whereas species that spend their entire life cycle in freshwater are typical of the Mamberamo region.

Although additional surveys are required before the overall picture is clear, the Cyclops coast may well be the best example of steep-gradient coastal stream habitat on the entire mainland of Papua. The only comparable areas that the author has surveyed occur on high offshore islands including Yapen, Biak, and islands of the Raja Ampat Group. This unusual and restricted freshwater ecosystem remains in excellent condition in the Yongsu area, lending support for the overall conservation significance of the Cyclops coast.

Most of the Yongsu fishes are widespread species that apparently possess a pelagic egg and/or pelagic larval stage that facilitates their dispersal. At present there are two freshwater species known only from the Cyclops coast, both in the sicydiine goby genus *Lentipes*. One of these, *Lentipes multiradiatus,* was collected for the first time during the current survey (Allen, 2001). The other, *Lentipes dimetrodon,* is known only from Omamerwai Creek, a small waterway about 12 km west of Yongsu (Watson and Allen, 1999). Further surveys, especially in the area between Vanimo, PNG, and the mouth of the Mamberamo River, are required to clearly delineate the distribution of these species. If they ultimately prove to be restricted to the Cyclops area it would be prudent to initiate conservation measures to ensure their protection.

The Dabra area probably does not possess any locally endemic species that would warrant urgent conservation measures. A possible exception is the diminutive Tiger Goby (*Eugnathogobius tigrellus*), but further collecting will probably show that this small and inconspicuous species has a much wider distribution than is currently recognised. On the other hand the Mamberamo Basin contains the highest percentage of endemic fish (21.4) of all major New Guinea rivers (Table 7.1). Although five species are endemic to the combined Sepik-Ramu systems, relatively few are unique to only one of these rivers. Of particular conservation concern is that this extraordinary endemism is nearly matched by an exceptionally high percentage (17.1) of introduced species, surpassing that of all other New Guinea rivers. Exotic fishes were introduced mainly during the 1970s and 1980s, presumably for food, and now appear to be well established.

Virtually all of the species introduced to the Mamberamo are known to negatively impact native fish populations wherever they have been released. They compete for living space and available food resources, or by feeding directly on native species. Tilapia (*Oreochromis mossambica*) and carp (*Cyprinus carpio*) are also notorious for creating turbid conditions in formerly clean lakes, and frequently displace native fishes due to their prolific breeding. It is probably too late to eradicate these species from the Mamberamo system, but every effort should be made to prevent further introductions into this or other river systems in Papua. Developing a practical solution to these problems appears to be a long way off, and no concern about this issue has been expressed by government agencies.

Additional ichthyological surveys in the Mamberamo Basin are urgently required. In terms of adding new records to the fauna and elucidating distributional patterns of the known fauna, future surveys should focus on two areas: 1. Mamberamo Delta, including Lake Rombebai, and 2. the major foothill tributaries around the edge of the Mamberamo Basin.

LITERATURE CITED

Allen, G.R. 1991. Field Guide to the Freshwater Fishes of New Guinea. Madang: Christensen Research Institute.

Allen, G.R. 2001. *Lentipes multiradiatus,* a new species of freshwater goby (Gobiidae) from Irian Jaya, Indonesia. aqua, Journal of Icthyology and Aquatic Biology 4: 121–124.

Allen, G.R. and M. Boeseman. 1982. A collection of freshwater fishes from western New Guinea with descriptions of two new species. Rec. West. Aust. Mus. 10(2): 67–103.

Allen, G.R. and D. Coates. 1990. An ichthyological survey of the Sepik River system, Papua New Guinea. Rec. West. Aust. Mus. (suppl). 34: 31–116.

Allen, G.R., L.R. Parenti, and D. Coates. 1992. Fishes of the Ramu River, Papua New Guinea. Ichthyol. Expl. Freshw. 3(4): 289–304.

Collette, B.C. 1982. Two new species of freshwater halfbeaks (Pisces: Hemiramphidae) of the genus *Zenarchopterus* from New Guinea. Copeia 1982(2): 365–276.

Haines, A.K. 1979. Purari River (Wabo) Hydroelectric Scheme. Environmental Studies Vol. 6 - An ecological survey of fish of the Lower Purari River system, Papua New Guinea. Port Moresby: Office of Environment and Conservation and Department of Minerals and Energy.

Löffler, E. 1977. Geomorphology of Papua New Guinea. Canberra, Commonwealth Scientific and Industrial Research Organization.

Parenti, L.R., and G.R. Allen. 1991. Fishes of the Gogol River and other coastal habitats, Madang Province, Papua New Guinea. Ichthyol. Expl. Freshw. 1(4): 307–320.

Roberts, T.R. 1978. An ichthyological survey of the Fly River in Papua New Guinea with descriptions of new species. Smiths. Contr. Zool. 281: 1–72.

Watson, R.E. and G.R. Allen. 1999. New species of freshwater gobies from Irian Jaya, Indonesia (Teleostei: Gobioidei: Sicydiinae). Aqua, J. Ichthyol. Aquat. Biol. 3(3): 113–118.

Weber, M. 1913. Susswasserfische aus Niederlandissh Sud- und Nord Neu Guinea. Nova Guinea (Leiden). 9(4): 513–613.

Weber, M. and L.F. de Beaufort. 1911–1962. The fishes of the Indo-Australian Archipelago. Volumes I–XI. Leiden: E.J. Brill.

Chapter 8

Amphibians and Reptiles of the Yongsu area, Papua, Indonesia

*Stephen Richards, Djoko T. Iskandar,
Burhan Tjaturadi, and Aditya Krishar*

CHAPTER SUMMARY

- 26 species of reptiles and 8 species of frogs were recorded in the vicinity of Yongsu on the northern edge of the Cyclops Mountains between 19–29 August, 2000.

- Two species of frogs and one species of lizard appear to be new to science.

- Marine turtles nest on beaches in the Yongsu area and are consumed by local communities. Their use and status should be monitored to determine whether current levels of utilisation are sustainable.

RINGKASAN BAB – HERPETOFAUNA YONGSU

- 26 spesies reptilia dan 8 spesies katak berhasil ditemukan di sekitar Yongsu, sebelah utara Pegunungan Cyclops antara 19–29 Agustus 2000.

- Dua spesies katak dan satu spesies kadal merupakan spesies baru bagi ilmu pengetahuan.

- Dua spesies penyu laut bertelur di kawasan pantai Yongsu dan dikonsumsi oleh penduduk setempat. Pemanfaatandan statusnya perlu di pantau untuk menentukan a pakah tingkat pendanfaatannya sudah lestari.

INTRODUCTION

The herpetofauna of Papua includes reptiles (snakes, lizards, turtles, and crocodiles) and frogs. Salamanders and caecilians, which together with frogs make up the class Amphibia, are absent from New Guinea. Surveys for these taxa in Papua have lagged behind neighboring Papua New Guinea, and there have been no serious attempts to survey the herpetofauna of the Cyclops Mountains since Evelyn Cheesman collected there more than 60 years ago.

A preliminary survey of herpetofauna in the Yongsu region on the northern coastal fringe of the Cyclops Mountains in 1999 documented seven species of frogs and reptiles (Conservation International, unpubl.). The current survey aimed to make a more detailed inventory of frogs and reptiles in the Yongsu region and to train Papuan biologists in biodiversity assessment techniques for this group of organisms.

METHODS

Our survey was concentrated along four trail networks in two small areas of hill forest near Yongsu Dosoyo. In both areas we accessed altitudes between 0 and 400 m asl. Trails 1 and 2 were adjacent to Yemang training camp, trail 3 was in Jari forest, and trail 4 was south of Yongsu Dosoyo:

Trail 1: Running north from Yemang Camp for about 100 m through lowland forest before ascending steeply to the west, reaching 400 m on a ridge west of camp. The trail loops to the east, descending steeply through hill forest and old gardens, re-entering the southern end of the camp.

Trail 2: Exits due south of Yemang camp, climbing steeply from the southern edge of camp to about 200 m a.s.l., and continues in a SSE direction through undulating hill forest.

Trail 3: Jari Forest. Runs through primary hill forest on a ridge across the bay from Yemang camp.

Trail 4: Extends from Yongsu Dosoyo southwards through heavily disturbed lowland forest and gardens to a waterfall at the base of the Cyclops Mountains.

All trails were sampled at night, but only trails 1, 2, and 3 were sampled during the day. The Yemang camp site was considered part of trail 1. During each sampling period we searched along trails for two to four hours. During the day we collected reptiles by hand from beneath rocks and logs, and from the forest floor and low trees. Small, active lizards were stunned with large rubber bands prior to capture. At night we conducted visual searches for herpetofauna by walking slowly along trails with headlamps; we also conducted audio searches for frogs and recorded the advertisement calls of frogs detected in this manner. We established 15 litter plots (5 m X 5 m) on trails 2 and 3 by randomly selecting plot sites with the use of a random number table. Four persons searched each plot carefully, one person on each side of the plot working towards the center. Because capture rates were low we combined frogs and reptiles from both trails for analysis. To determine local names for species recorded during this survey we interviewed two local community members, Pak Ilius and Pak Simpson, who were recommended as having the greatest knowledge about local herpetofauna.

RESULTS

Twenty-six species of reptiles and eight species of frogs were recorded from the Yongsu area (Appendix 26). An additional five reptiles and two frogs were recorded during two previous training courses (Table 8.1). Two species of frogs are new to science. One of these (*Hylophorbus* sp.) is currently being described by Dr. R. Günther of the Berlin Museum. The other (*Oreophryne* sp.) will be described within the next 12 months by the senior authors. Another frog heard calling from high in the forest canopy almost certainly represents an undescribed species of *Litoria*. Further attempts should be made to collect and identify this potentially very interesting canopy-dwelling species.

Five species (two frogs, three reptiles) were recorded in litter plots. The species accumulation curve for litter herpetofauna reached an asymptote after just eight plots (Figure 8.1). General searching of the forest floor along trails 1–3 demonstrated that the litter herpetofauna is substantially more diverse than indicated by these plots. Considerable time and effort are required to establish and search plots, and this technique did not reveal any species that were not detected during general surveys. Litter plots are not particularly useful for rapid, comprehensive inventories of New Guinea litter herpetofauna.

Two species of marine turtle nest on the beaches at Yongsu: the green turtle (*Chelonia mydas*) and the loggerhead turtle (*Caretta caretta*). Green sea turtle hatchlings emerged

Table 8.1. Additional frogs and reptiles recorded during previous Yongsu training courses (1999–2000).

Amphibia
Microhylidae
Cophixalus cheesmanae?
Sphenophryne cornuta
Reptilia
Scincidae
Tribolonotus gracilis
Boidae
Candoia carinata
Morelia amethistina
Colubridae
Stegonotus modestus
Elapidae
Acanthophis antarcticus

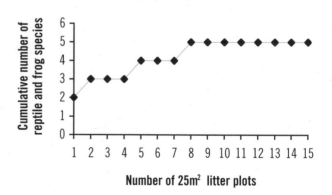

Figure 8.1. Species accumulation curve for litter herpetofauna at Yongsu, Papua.

Table 8.2 Local names for Herpetofauna at Yongsu.

Local name	Scientific name	Comments
Snakes		
Bufnotu	*Candoia aspera*	"Water snake"; refers to *Candoia* only when found in water
Angtu	*Candoia aspera*	Refers to this species when found on the ground
Mereng	*Morelia amethistina*	
Owarre	*Tropidonophis*	
Awedothi-Dothi	*Boiga irregularis*	
Mereng	*Stegonotus*	All species of this genus, including different color patterns
Onyariwaij	*Dendrelaphis, Morelia viridis*	Tree-dwelling snakes
Angthu	*Micropechis ikaheka*	Black-headed snakes
Dokhrew	*Acanthophis antarcticus*	
Lizards		
Kipauway	*Emoia caeruleocauda, Emoia jakati*	All striped skinks
Pangkhoru	*Sphenomorphus/Emoia*	All brown skinks
Bureng	*Lamprolepis smaragdina*	
Arehka	*Hypsilurus* sp. nr *dilophus*	
Firakho	*Hypsilurus modestus*	
Besereway	*Cyrtodactylus/Gehyra*	Geckos
Prong	*Varanus jobiensis*	
Frogs		
Sikha	Frogs	All frogs

from a nest at Yongsu camp at 6:30 p.m. on 25 August 2000 and a single loggerhead turtle hatchling was observed at the same locality on 28 August 2000.

Local names for reptiles and frogs are presented in Table 8.2. Some names are applied to ecological or morphological groups of species rather than to "biological" species as recognised by western science. Conversely in some instances one species may have more than one name depending on the habitat in which it is observed.

DISCUSSION AND CONCLUSIONS

The herpetofauna of the Yongsu area is largely an extension of the widespread lowland rainforest fauna of northern New Guinea. The most abundant species including *Platymantis papuensis*, *Sphenomorphus simus*, *Emoia caeruleocauda*, and *E. jakati* are widespread in New Guinea, as are most of the other species that we were able to identify with any certainty. Endemic Cyclops species such as *Platymantis cheesmani* were not encountered. One species of *Hypsilurus* that possibly represents an undescribed species, and the undescribed *Oreophryne* may be locally restricted Cyclops Mountains

taxa, but further studies are required to determine their status and distribution.

Snakes were poorly represented in our survey, and further studies might be expected to increase the known diversity of this group. A single sea snake (*Laticauda* sp.) was observed by Gerald R. Allen on the reef off Yongsu, and several other marine snakes might occur in the area. However, with the exception of the large canopy-dwelling *Litoria*, few additional frogs are likely to be found in the area. The general paucity of stream-dwelling frogs was surprising and may reflect the fact that streams in the Yongsu area are very short, rising from the steep ridges close to the sea.

Forests in the Yongsu area remain largely intact and provide habitat for a herpetofauna representative of primary rainforest in northern New Guinea. Although the Yongsu area is outside the Cyclops Nature Reserve, the opportunity exists to work with local communities to preserve this lowland and foothill forest. We saw no evidence that any activities by local communities are negatively impacting terrestrial herpetofauna. However, impacts of local communities on marine turtles are less clear. Future foci of conservation activities in the area should include monitoring of nesting activity and hatching success of marine turtles, and documentation of turtle harvesting by local communities.

Chapter 9

Amphibians and Reptiles of the Dabra area, Mamberamo River Basin, Papua, Indonesia

Stephen Richards, Djoko T. Iskandar, and Burhan Tjaturadi

CHAPTER SUMMARY

- Twenty-one species of frogs and 36 species of reptiles were recorded from three sites in the Mamberamo River Basin (Furu River, Tiri River, and Dabra Village).

- Seven species of frogs and up to three species of lizards are unknown to science.

- Documentation of the Sail-fin Lizard *Hydrosaurus amboinensis* at Furu and Tiri represents a significant easterly range extension for this large and spectacular reptile.

- Several large reptiles including the Freshwater Crocodile (*Crocodylus novaeguineae*) and two freshwater turtles—the Softshell Turtle (*Pelochelys cantori*) and the New Guinea Side-Neck (*Elseya novaeguineae*)—are harvested by local communities and represent a source of cash income and food.

RINGKASAN BAB – HERPETOFAUNA DABRA

- 21 spesies katak dan 36 spesies reptilia berhasil dicatat dari tiga lokasi di daerah aliran Sungai Mamberamo (Sungai Furu, Sungai Tiri, dan Kampung Dabra).

- Ditemukan 7 spesies katak dan 3 spesies kadal yang belum diketahui dunia ilmu pengetahuan.

- Dokumentasi Kadal Sirip Layar *Hydrosaurus amboinensis* di Furu dan Tiri menunjukkan perluasan jelajah kearah timur yang signifikan untuk reptilia besar dan spektakuler ini.

- Beberapa reptilia besar, termasuk Buaya Air tawar (*Crocodylus novaeguineae*) dan dua kura-kura air tawar – Labi-labi Raksasa (*Pelochelys cantori*) dan Kura-kura Irian (*Elseya novaeguineae*) – di tangkap oleh masyarakat sebagai sumber pendapatan dan makanan.

INTRODUCTION

Early attempts to document Papua's herpetofauna concentrated on islands off the western and northern coasts of the mainland (as described in Richards et al., 2000). Important collections were also made on the southern slopes of the Snow Mountains, and in the Arfak Mountains (e.g. Boulenger, 1914; Peters and Doria, 1878). Few herpetological collections have been made in the Mamberamo River Basin, a vast area of lowland rainforest and swamps on the northern side of New Guinea's central cordillera. The 1938–1939 Archbold Expedition collected frogs and reptiles in the Mamberamo area but most survey effort was concentrated at altitudes ≥ 850 m a.s.l. and specimens were collected opportunistically by scientists specializing in other taxa (Archbold et al., 1942). As a result relatively few frog and reptile specimens were collected compared to other taxa and the herpetofauna of the lowland rainforests was poorly documented. Despite this lack of survey effort the region is known to harbor many interesting species including an endemic genus of aquatic snake (O'Shea, 1996). Participants at the Irian Jaya Biodiversity Conservation Priority-Setting Workshop held in Biak in 1997 recognized that the area has a diverse herpetofauna and that further research into the fauna is a very high priority (Conservation International, 1999).

The aims of this survey were to quantify herpetofaunal diversity in the Dabra region of the Mamberamo River Basin and to provide field-based training for an Indonesian herpetologist, Mr. Burhan Tjaturadi.

METHODS

We surveyed frogs and reptiles at three localities; Furu, Tiri, and Dabra Village. The survey schedule is outlined in Table 9.1. At each site we identified all accessible habitats and searched them thoroughly during the day and at night. During each sampling period we conducted searches for two to four hours. At night we conducted visual searches for frogs, geckos, and snakes by walking slowly along forest trails, streams, and through swamps with headlamps. We also conducted audio searches for frogs and recorded the advertisement calls of frogs detected in this manner. During the day we conducted visual transects for diurnal reptiles. Small lizards were caught by hand or stunned with a large rubber band prior to capture. Large snakes and lizards were caught by hand, and one participant (SJR) snorkeled for turtles. We interviewed our guides, Klaus and Marcus, to obtain local (Dasigo) names for frogs and reptiles. Some identifications, particularly of small lizards, are tentative and may change following careful examination of taxonomically difficult taxa. A species accumulation curve was constructed using only days when at least two to four hours of survey effort was undertaken during the day and at night.

Table 9.1. Schedule (inclusive days) for herpetofauna survey in the Mamberamo area.

Site	Dates	# of personnel
Furu	1–7 September	5
Tiri	8–12 September	5
Dabra	13–14 September	2

RESULTS

A total of 21 species of frogs and 36 species of reptiles were recorded from the three sites (Appendix 27). At least seven frogs appear to be undescribed species. Microhylids and ranids dominated the frog fauna but three of the seven undescribed frogs were hylids (tree frogs), a group that was otherwise poorly represented at Dabra. An undescribed species of *Oreophryne* (Microhylidae) is only the third frog species in the world known to guard direct-developing eggs exposed on leaves in the forest (Myers, 1969; Johnston and Richards, 1993).

The Freshwater Crocodile (*Crocodylus novaeguineae*) and two freshwater turtles, the Giant Softshell Turtle (*Pelochelys cantori*) and the New Guinea Side-Neck (*Elseya novaeguineae*), were harvested by local communities at the Tiri site. Only one of these species, *Elseya novaeguineae*, occurred at the Furu site. Although crocodile populations in the area appear to have decreased due to intensive hunting for skins, turtles were relatively abundant—two *Pelochelys* were recorded during just five days at the Tiri site. Other species used as food by local communities included two species of monitors (*Varanus jobiensis* and *V. indicus*). Our records of the Giant Sail-Fin Lizard (*Hydrosaurus amboinensis*) appear to be the first for this large and spectacular animal east of the Vogelkop Peninsula on mainland New Guinea. An overview of noteworthy species documented during this survey is presented in Appendix 28.

A species accumulation curve (Figure 9.1) shows that reptile diversity in the Dabra region is likely to be substantially higher than we documented, but that relatively few additional frogs would be recorded with additional effort under similar conditions. However, weather conditions were quite dry during our survey, and it is likely that additional surveys during the wet season will increase the known frog fauna

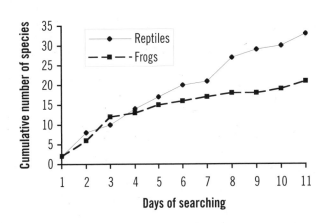

Figure 9.1. Species accumulation curve for Herpetofauna at Mamberamo sites combined.

substantially. Local (Dasigo) names for frogs and reptiles are presented in Table 9.2.

DISCUSSION AND CONCLUSIONS

This survey documented an impressive diversity of herpetofauna in the Dabra region of the Mamberamo River Basin. Reptile and frog species were added to the inventory until the last day of the survey, but the species accumulation curve indicates that few additional frogs are likely to be discovered during the dry season using the same survey techniques. Increases in our species inventory were most apparent when we changed sites, reflecting access to slightly different habitats. This suggests that access to different areas in the Mamberamo catchment and to a wider range of altitudes (and therefore forest types) will increase the known herpetofaunal diversity even further. During our survey the Dabra region had little rainfall, rivers were low, and frog activity was minimal. Further investigations in the general area during the wet season will certainly reveal many additional frog species. The high proportion of undescribed frogs is not surprising (Richards et al., 2000) and indicates how poorly known the fauna of this region is. It is likely that the lowland rainforests of the Mamberamo River Basin harbor an exceptionally diverse herpetofauna that will greatly exceed the total documented during our geographically limited survey.

One of the most significant outcomes of this survey was to document the use of reptile species by the Dabra community. Crocodile populations have been over-hunted in the area, but at least two species of freshwater turtles appeared to be abundant despite being hunted regularly for food. Any future study of resource economics in this region should take into account the value of reptiles as a food source for local communities, and the conservation status and current intensity of hunting of these species should be assessed as a matter of high priority.

The Dabra herpetofauna is substantially more diverse than that recorded at Yongsu (Chapter 8, this volume), reflecting a more completely developed lowland rainforest and a wider range of habitats. Several widespread taxa typical of lowland rainforests, including *Platymantis papuensis, Hypsilurus modestus,* and *Stegonotus* spp., were common to both areas. However, there were a large number of species recorded at Dabra that were not documented from Yongsu. Interestingly the fauna at Yongsu was not entirely a sub-set of the more diverse Mamberamo fauna, but contained at least two species (a *Hypsilurus* sp. with blue eyes and an *Oreophryne* sp.) that were not collected at Dabra and are not known from elsewhere in the northern lowlands. Whether these species represent a lowland extension of an endemic Cyclops Mountains fauna or simply have yet to be documented elsewhere remains to be determined.

Table 9.2. Dasigo names for herpetofauna in the Dabra area, Papua, Indonesia.

Common/scientific name	Local (Dasigo) name
Frogs	
Rana arfaki	Fro
All brown frogs except *Rana arfaki*	Feidogukij
All green *Litoria* species	Feidoguu
Lizards	
Cyrtodactylus mimikanus	Feidaurij
House geckos	Kwariwidjda
All brown *Hypsilurus* sp.	Kei
Hypsilurus modestus	Fujde
Hydrosaurus amboinensis	Tagu
All monitors (*Varanus*)	Kwesij
All skinks	Kurigi
Turtles	
Elseya novaeguineae	Kaij
Pelochelys cantori	Fruda
Snakes	
Acanthophis antarcticus	Tbifga
Aspidomorphus muelleri	Tbisiri
Stegonotus species	Tbisiri
Tropidonophis montanus	Tbisiri
Crocodiles	
Crocodylus novaeguineae	Feijda

LITERATURE CITED

Archbold, R., A.L. Rand, and L.J. Brass. 1942. Results of the Archbold Expeditions No. 41. Summary of the 1938–1939 New Guinea expedition. Bull. Am. Mus. Nat. Hist. 79: 197–288.

Boulenger, G.A. 1914. An annotated list of the batrachians and reptiles collected by the British Ornithologists' Union Expedition and the Wollaston Expedition in Dutch New Guinea. Trans. Zool. Soc. Lond. 20: 247–274.

Conservation International. 1999. The Irian Jaya Biodiversity Conservation Priority-Setting Workshop: Final report. Washington, DC: Conservation International.

Johnston, G.R. and S.J. Richards. 1993. Observations on the breeding biology of a microhylid frog (Genus *Oreophryne*) from New Guinea. Trans. R. Soc. S. Aust. 117: 105–107.

O'Shea, M.T. 1996. A guide to the snakes of Papua New Guinea. Port Moresby: Independent Publishing.

Myers, C.W. 1969. The ecological geography of cloud forest in Panama. Am. Mus. Nov. 2396: 1–52.

Peters, W. and G. Doria. 1878. Catalogo dei rettili e dei batraci raccolti da O. Beccari, L.M. D'Albertis e A.A. Bruijn nella sotto-region e Austro-Malese. Ann. Mus. Civ. Stor. Nat. Genova. 13: 323–450.

Richards, S.J., D.T. Iskandar, and A. Allison. 2000. Amphibians and reptiles of the Wapoga River area, Irian Jaya, Indonesia. *In*: A. Mack and L. Alonso (eds.). A biological assessment of the Wapoga River area of northwestern Irian Jaya, Indonesia. RAP Bulletin of Biological Assessment 14. Washington, DC: Conservation International. Pp. 54–57.

Chapter 10

Birds of the Yongsu area, northern Cyclops Mountains, Papua, Indonesia

Pujo Setio, Paul Johan Kawatu, David Kalo, Daud Womsiwor, and Bruce M. Beehler

CHAPTER SUMMARY

- Ninety species of birds were recorded during ten days of observations in the Yongsu Dosoyo area; most of these were forest-dwelling birds. Lack of a coastal plain and lack of significant lowland forest are factors that probably contribute to the relatively species-poor bird fauna.

- Fifteen birds typical of lowland forest, including the Victoria Crowned Pigeon and the Common Paradise-Kingfisher, were apparently absent. Conversely, the presence of healthy populations of Blyth's Hornbill, Palm Cockatoo, and Northern Cassowary indicate that this forest is not depleted of its megafauna.

- Point counts indicated that the forest avifauna is quite patchy at a local level. Informal censuses of forest trees indicate that the arboreal forest flora is patchily distributed. There was no evidence that small-scale selective logging in this forest has harmed local bird populations.

- Reports from a local naturalist provided evidence that the Cyclops Mountains remain significantly under-surveyed for birds and mammals.

RINGKASAN BAB – BURUNG-BURUNG YONGSU

- 90 spesies burung berhasil dicatat selama 10 hari pengamatan di daerah Yongsu-Dosoyo; kebanyakan adalah burung-burung hutan. Tidak adanya dataran pantai dan tidak adanya hutan dataran rendah yang signifikan merupakan faktor yang mengakibatkan sedikitnya jumlah spesies burung.

- Lima belas spesies burung tipikal hutan dataran rendah, termasuk Mambruk Victoria dan Cekakak-Pita Biasa tidak dijumpai. Sebaliknya, adanya populasi yang sehat dari Rangkong Papua, Kakatua Raja, dan Kasuari Gelambir Satu menunjukkan bahwa di hutan ini tidak terjadi penurunan jumlah fauna besar.

- *Point counts* (teknik penghitungan pada suatu titik) mengindikasikan bahwa sebaran burung-burung hutan kurang merata di tingkat lokal. Sensus informal pohon-pohon hutan mengindikasikan bahwa tumbuh-tumbuhan arboreal juga tersebar tidak merata. Tidak ada bukti bahwa pembalakan kayu skala kecil yang selektif di hutan ini telah membahayakan populasi burung.

- Laporan-laporan dari naturalis lokal memberikan bukti belum banyaknya survei burung dan mamalia di Pegunungan Cyclops

INTRODUCTION

The bird fauna of Papua is dominated by forest species, many of which are widespread lowland forms (Beehler et al., 1986; Conservation International, 1999). Although lowland forest is the most accessible habitat on the island of New Guinea, the avifauna of this ecosystem remains poorly documented. Even a brief survey of birds around Yongsu Dosoyo can provide useful new information relevant to biodiversity studies of the lowland rainforest environment of New Guinea.

Yongsu is of particular interest because of its peculiar physiographic location in a coastal zone with high topographic relief and no coastal plain. The mountain ridges plunge from

Table 10.1. Results of point censuses for birds by two census teams in Yongsu forest, Papua, Indonesia.

Census Team A			
CENSUS POINT 3			
DATE	TIME	# SPECIES	# INDIVIDUALS
24-Aug	9:22	12	22
25-Aug	9:00	13	27
26-Aug	8:35	19	36
28-Aug	8:05	16	40
	MEANS	**14.8**	**30.4**
CENSUS POINT 4			
DATE	TIME	# SPECIES	# INDIVIDUALS
24-Aug	9:40	12	26
25-Aug	9:30	9	19
26-Aug	8:11	12	20
28-Aug	8:30	17	36
	MEANS	**12.6**	**26.6**
Census Team B			
CENSUS POINT 3			
DATE	TIME	# SPECIES	# INDIVIDUALS
24-Aug	9:37	21	32
25-Aug	9:30	14	29
26-Aug	8:11	19	26
28-Aug	9:20	21	29
28-Aug	8:35	24	32
	MEANS	**19.8**	**29.6**
CENSUS POINT 4			
DATE	TIME	# SPECIES	# INDIVIDUALS
24-Aug	9:16	16	19
25-Aug	9:07	10	26
26-Aug	8:32	8	14
28-Aug	8:12	21	25
28-Aug	8:58	16	28
	MEANS	**14.2**	**22.4**

more than 1000 m elevation to sea level in a distance of no more than 5 km. The aim of this survey was to document bird diversity within the narrow strip of coastal lowland forest abutting the northern edge of the Cyclops Mountains.

METHODS AND STUDY SITE

Birds were documented in a tract of selectively logged hill forest on a dissected and sloping narrow plateau (40–70 m asl) about 750 m south of the Yemang training camp. See Chapters 1 and 2 of this volume for descriptions of the habitat. The survey was conducted over a ten-day period (20–29 August 2000) using mist-netting, audial censuses, point counts, and *ad lib* observations. A few additional *ad lib* observations were made at other sites within a few kilometers of the camp. Rain rarely interrupted field work so conditions for observing and netting birds were excellent. An array of four census points was plotted and mapped, but time constraints forced our team to concentrate on censuses at two points (3 and 4) to allow sufficient practice in point-census techniques and at the same time to compare avian use of forest at two widely separated (250 m) census points. Mist nets were situated in and around the forest at census point 3.

RESULTS AND DISCUSSION

We recorded 90 species of birds during 50 net-days, 240 minutes of point censuses, and about 120 hours of *ad lib* and transect audial surveys (Appendix 29). These included a range of large species usually desired by hunters for game and plumes (Northern Cassowary, Brown-collared Brush Turkey, Blyth's Hornbill, Palm Cockatoo, White-bellied Sea-Eagle, Lesser Bird of Paradise, Magnificent Riflebird). Five species of birds of paradise were recorded. Species accumulation followed a typical pattern (cf. Beehler et al., 1995), and the lack of evidence of an asymptote (Figure 10.1) indicates that our survey effort was incomplete. Beehler & Mack (1999) have shown that for New Guinea forest avifaunas, a minimum of two months is required to generate a relatively complete species list. Another indication of the incompleteness of this survey is that 13 species were recorded only once during the ten-day effort.

A comparison with other New Guinea lowland forest bird surveys generated from ten days of effort (see Beehler et al. 1995, Figure 3) shows that the Yongsu list is perhaps 20 to 30 species lower than expected.

The point census results (Table 10.1) indicate that this technique is of limited use for generating a comprehensive species list. A maximum of 24 and a minimum of 8 species were recorded during these censuses. Nonetheless, the point counts did highlight considerable inter-site patchiness in forest use by birds, even between points in moderately close proximity (250 m). For example, at census point 3 Team B recorded Blyth's Hornbill on all six censuses (for a total of

12 sightings), whereas at point 4 this species was recorded by Team B on only two censuses, for a total of three birds. The New Guinea Longbill exhibited a similar pattern: point 3—seven individuals recorded on six censuses *vs.* point 4—one individual recorded on one census. Clearly birds are not equitably distributed through the forest, even at the micro-patch scale. An informal census of mature forest trees at our census sites indicates considerable patchiness of tree species at the local level (Appendix 30). Might forest tree patchiness influence bird patchiness? More study will be required to test this.

Similarly, mist-netting conducted with ten nets over five days trapped just 29 individuals of 18 species. This accords with Beehler and Mack's (1999) assertion that mist-netting in New Guinea is a time-consuming method that is valuable mainly for specialized studies, rather than for general biotic surveys. At Yongsu many bird species were identified solely on the basis of their vocalizations, indicating that effective short-term surveys will require field staff with excellent knowledge of local bird song.

One of the most surprising results of this survey is that a number of widespread lowland forest bird species were not recorded at Yongsu (Table 10.2). These included large or conspicuous birds like the Victoria Crowned Pigeon, Common Paradise Kingfisher, Hooded and Spot-wing Monarchs, and Twelve-wired Bird of Paradise. These species were unlikely to be overlooked during our survey and were not reported from the area by local informants. It appears that the local forest avifauna is impoverished, and the most

Table 10.2. Widespread lowland forest birds apparently missing from Yongsu forest.

SPECIES
Trugon terrestris
Goura victoria
Chalcopsitta duivenbodei
Psittaculirostris edwardsi
Eudynamys scolopacea
Chaetura novaeguineae
Tanysiptera galatea
Monarcha manadensis
Monarcha guttula
Poecilodryas hypoleuca
Lichenostomus obscurus
Oriolus szalayi
Peltops blainvillii
Ailuroedus buccodes
Seleucidis melanoleuca

obvious missing species are those typical of lowland alluvial forest.

Several take-home points can be made from our brief ornithological field effort at Yongsu:

1. It appears that lowland bird faunas are impoverished at sites where rugged mountains descend to the coast, preventing the development of a coastal plain. There is little evidence that this absence of key lowland species is compensated for by the infusion of hill or montane forms.

2. Point census efforts are valuable for highlighting local patchiness of forest birds, which is a significant and little-studied phenomenon in lowland rainforests (Beehler and Mack, 1999).

3. The presence of apparently healthy populations of cassowaries, cockatoos, brush-turkeys, and birds of paradise in lightly logged forest in the immediate vicinity of an active village indicates that under certain circumstances long-term management of forest for a range of timber and nontimber products may be environmentally sustainable. Forest use by the Yongsu communities should be studied further to determine how they have managed to extract resources without seriously impacting the forest megafauna. There may be valuable lessons to be learned from their subsistence practices.

4. Reports by a knowledgeable local informant indicate that the forests of the northern Cyclops Mountains may harbor undocumented species of considerable taxonomic significance. Intensive field studies of the upland biota of the northern Cyclops are strongly recommended.

LITERATURE CITED

Beehler, B. and A. Mack. 1999. Constraints to characterising spatial heterogeneity in a lowland forest avifauna in New Guinea. *In:* Brawn, J. and S. Robinson (eds.): Symposium on Avian Community Ecology in Tropical Forests. Durban: 22nd Internat. Ornith. Congress. Pp. 2569–2579.

Beehler, B., T.K. Pratt, and D.A. Zimmerman. 1986. Birds of New Guinea. New Jersey: Princeton University Press.

Beehler, B., J. Sengo, C. Filardi, and K. Merg. 1995. Documenting the lowland rainforest avifauna in Papua New Guinea—effects of patchy distributions, survey effort, and methodology. Emu. 95: 149–161.

Conservation International. 1999. The Irian Jaya Biodiversity Conservation Priority-Setting Workshop: Final report. Washington, DC: Conservation International.

Chapter 11

Birds of the Dabra area, Mamberamo River Basin, Papua, Indonesia

Bas van Balen, Suer Suryadi, and David Kalo

CHAPTER SUMMARY

- One hundred forty-three species of birds were recorded from lowland forest, swamps and riverine habitats in the Dabra region; of these 65 (45%) are endemic to New Guinea.

- Birds of prey (11 species) and birds of paradise (6 species) were especially well represented.

- Two threatened species (Northern Cassowary and Victoria Crowned Pigeon) that are vulnerable to hunting were relatively common in the area, indicating that hunting pressure from local communities is low.

- Swamp forests to the north of Dabra should be surveyed for rare local endemics not found during the present survey, such as the Pale-billed Sicklebill and Brass's Friarbird.

RINGKASAN BAB – BURUNG-BURUNG DABRA

- 143 spesies burung tercatat keberadaannya mulai dari hutan dataran rendah, rawa-rawa, dan habitat riparian (sekitar sungai) di daerah Dabra; sebanyak 65 spesies (45%) dari jumlah tersebut adalah endemik New Guinea.

- Burung-burung pemangsa (11 spesies) dan burung-burung Cenderawasih (6 spesies) terwakili dengan baik.

- Dua spesies terancam (Kasuari Gelambir-tunggal dan Mambruk Victoria) yang rentan terhadap perburuan, secara relatif umum dijumpai di daerah ini, mengindikasikan bahwa tekanan perburuan dari masyarakat lokal tergolong rendah.

- Hutan rawa di sebelah utara Dabra perlu disurvei untuk menemukan burung endemik lokal langka yang tidak ditemukan dalam survei ini, misalnya Paruh-sabit Paruh-putih, *Epimachus bruijnii* dan Cikukua Mamberamo, *Philemon brassi.*

INTRODUCTION

The Mamberamo region covers a vast area of diverse and pristine ecosystems incorporating montane forest (up to 2350m in the Van Rees Mts), hill and lowland rainforest, swamp forest, swamps and other freshwater systems, and estuarine habitats. Lack of accessibility and low population density have protected the basin from many extractive industries that have damaged forests elsewhere in New Guinea. The Irian Jaya Biodiversity Conservation Priority-Setting Workshop (Conservation International, 1999) concluded that the Mamberamo Basin is an area with high priority for conservation, and is in urgent need of biological and ecological research.

The lower reaches of the Mamberamo-Idenburg-Rouffaer rivers are part of the North Papuan lowlands Endemic Bird Area (EBA), which harbors nine birds with restricted global ranges (Stattersfield et al. 1998). Three of these are endemic to the EBA: the threatened (vulnerable) Salvadori's Fig-parrot, *Psittaculirostis salvadorii;* the near-threatened Brass's Friarbird, *Philemon brassi;* and the near-threatened Pale-billed Sicklebill, *Epimachus bruijni.* Two Wildlife Sanctuaries, the Mamberamo-Foja and the Rouffaer River, protect up to 1.4 million hectares of forest in the basin and incorporate a variety of habitats between sea level and 2,000 m asl. The few ornithological surveys undertaken in the region (Stattersfield et al., 1998) documented 332 bird species, but the status and distribution of the avifauna remains largely unknown. The most important surveys in the Mamberamo Basin have been:

1) Van Heurn (June 1920–January 1921), who collected bird skins of 120 species near Pionier (Kasonaweja or Atuoi), Batavia, and Prauwen Bivouacs (Hartert, 1932; van Heurn, 1921a, b).

2) The Archbold Expedition (1938–1939) that obtained about 1000 bird skins of 164 species within six weeks at Bernhard Camp (Archbold et al., 1942).

3) J.P.K. van Eechoud (July–November 1939), who collected 250 skins of 87 bird species close to Pionier Bivouac (van Bemmel, 1947).

4) J.M. Diamond (1979), who made a survey along the Mamberamo river (Diamond, 1980).

STUDY SITES

The areas we surveyed near Dabra village were ca 8–10 km S/SE of Van Heurn's Prauwen Bivouac, ca 60 km SSE of Batavia Bivouac, ca 125 km SE of Pionier Bivouac, and ca 70 km WNW of Archbold's Bernhard Camp. The main sites surveyed were:

Dabra—A small village on the banks of the Idenburg/ Teritatu River, bordering patchily well-forested hills to the south and with flat floodplains and swamps to the north. A grassy landing strip, hill forest, gardens (with large remnant, dead trees), shrubs, and some paddy fields constituted the major habitats surveyed. Dabra is the main village in Mamberamo Hulu District.

Furu—The survey team's camp on the banks of the Furu River at the edge of a well-forested hill. Observations were conducted up to about 150 m asl on the hillside, in swampy lowland forest that extended to the banks of the Mamberamo River, and along a forested creek that drained into

the Mamberamo River. Four mist nets were erected on the hill and six were set in lowland forest near camp.

Tiri—The survey team's second campsite on the banks of the Tiri River. Observations were conducted mainly along a sandy creek close to the camp and along a trail that led to a salt well in the forest, several hundred meters from camp. Five mist nets were placed in the forest about 150 m from camp.

Buare—A lagoon 8 kilometers east of Dabra. The banks were overgrown with cane grass, scattered shrub, and patches of riverine forest. At low water levels, the banks offered foraging sites for various water birds.

SURVEY METHODS

Surveys were carried out by walking slowly along dry river beds, existing transects (constructed by other teams), and trails within a radius of up to about 1.6 km from the Furu and Tiri campsites. Observations were made during five days from sunrise to sunset, with breaks of several hours around noon. Birds were identified visually or by their vocalizations, and the number of species recorded each day was tallied. A species accumulation curve was constructed from daily totals. An additional survey was made towards Buare, and observations were also made around Dabra and along trails between Dabra and the camp sites. We classified the abundance of species into four different categories: *Abundant* (recorded regularly, in moderate to large numbers); *Common* (recorded regularly, in small numbers); *Uncommon* (recorded irregularly); and *Rare* (recorded once or twice).

One day (12 September) was spent at a single point in the forest at Tiri. Five-minute tallies of all species seen and/or heard within a radius of about 50 m were made during two periods in which bird activity was expected to be the highest: 0515–0910 hrs, and 1520–1745 hrs.

Mist nets erected at the Furu and Tiri sites were 12 m long, included a range of mesh sizes, and were opened for 4 to 5 days. Captured birds were weighed and measured (tarsus, wing, bill), eye color and any molt was recorded (see Ginn and Melville, 1983), tail feathers were clipped for future recognition, and photographs were taken.

Vocalizations were recorded with a Sony MZ R30 Mini-Disk recorder and Sony ECM-PB1C Parabola Microphone to assist later identification if the bird was not immediately recognized. A selection of recordings will be deposited in the British Library Sound Archive (London).

Nocturnal surveys were conducted in and around our camps between sunset and sunrise, when night birds (owls, nightjars, frogmouths) were active. The herpetology team recorded bird vocalizations between dusk and 1:00 a.m., so we restricted our own observations to the short predawn period when night-birds are most active.

Forest-dwelling peoples are well known to have an intimate knowledge of their environment, and often have the ability to distinguish between morphologically similar but biologically distinct birds. The gradual disappearance of many local languages and the imminent threat of losing unique vocabularies led Diamond and Bishop (1999) to argue that biologists should record as much biological data as possible from local peoples' traditional knowledge systems. We have made an attempt to record bird names in two local languages (Dasigo and Airo) through semi-structured interviews with our local guides and porters (mainly Messrs Wenan, Musa, Claus, and Frits). Additional names in the Kaowerawetj language spoken in the Pionier Bivouac area were obtained from van Bemmel (1947).

RESULTS AND DISCUSSION

One hundred and forty-three species of birds were recorded during the two-week survey (Appendix 31). This total includes eight species that require final confirmation of identification, for example by additional analyses of sound recordings made during the RAP. It excludes a number of species reported by local people but not seen or heard by us. Eighty-seven species were found during five days at the Furu site, and 87 species were found during five days at the Tiri site. The combined species total for the two sites was 113 species. The remaining species were documented at Dabra, Buarae, and other sites nearby. Fifteen birds representing eight species were captured in mist nets at Furu, and eight birds belonging to five species were taken in nets at Tiri. The most significant of these were the Blue Jewel-Babbler (*Ptilorrhoa caerulescens*) and Paradise Kingfisher (*Tanisyptera galatea*), which were otherwise not seen during the survey, though their calls were heard in the surrounding forest. Only one bird, an adult Paradise Kingfisher (*Tanysiptera galatea*), was recaptured. Local guides collected a Magnificent Fruit-Dove (*Ptilinopus magnificus*) and found a dead Azure Kingfisher (*Alcedo azurea*).

A two-week survey was insufficient to obtain an accurate estimate of the total avifauna in our study area. Ten species recorded by Van Eechoud (Hartert 1932) at Prauwen Bivouac, in the same general area, were not found during the present survey (Appendix 32). Some of these were water birds, which were covered only marginally during our surveys, or were represented only by a few skins in early collections and thus are probably rare or locally distributed in the area. A combined species accumulation curve for Furu, Buare, and Tiri (Figure 11.1) indicates that further survey effort would reveal considerably more species in this area. The total number likely to be attained (by extrapolation) seems strikingly close to the 164 species recorded by the exhaustive Archbold Expedition.

Our attempt to find the poorly known Brass's Friar-bird, *Philemon brassi*, collected by Archbold et al. (1942) at Bernhard Camp, and recently observed in the Wapoga River catchment (Mack et al., 2000), was unsuccessful. We

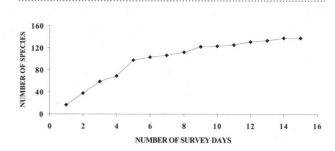

Figure 11.1. Species accumulation curve for birds at sites around Dabra, Mamberamo River basin.

spent just one morning at the Buare lagoon in habitat that appeared to be suitable for this species, and further surveys of nearby swampy areas with dispersed lagoons may confirm its presence. Other birds that may be present but difficult to detect because of patchy distributions or extremely low density include the Pale-Billed Sicklebill (*Epimachus bruijnii*), collected by both van Heurn (1921c) and van Eechoud (van Bemmel, 1947) near Pionier Bivouac, Brown-Headed Crow (*Corvus fuscicapillus*), and Rufous Whistler (*Pachycehala leucogastra*).

Diamond (1980) listed 199 species during his survey including 23 lowland species not seen by us. However the precise localities of his observations are unknown to us so his data are not included in Appendix 31.

Birds in tropical rainforest are difficult to observe, and most species are detected by their vocalizations. A thorough knowledge of the birds' vocal repertoires, including local dialects (Beehler and Mack, 1998), is therefore critical. Acquiring this knowledge for New Guinea birds takes many years, but once these skills are attained a comprehensive survey can be undertaken without the aid of nets and with only minimal reliance on visual identification of birds.

According to our guides, waterbirds are abundant in lakes around Mamberamo Hulu during the dry season (July to September). Many of these birds have apparently migrated from Australia. Unfortunately we did not observe large aggregations of waterbirds because the survey coincided with the end of their stay in New Guinea. We observed only a few Common Sandpipers *Actitis hypoleucos* (a northern migrant) and Black-winged Stilts (*Himantopus himantopus*) on the banks of the Mamberamo River and Buare lagoons.

Conditions during the last few days of the survey appeared to be ideal for night-bird activity (dry, moon), but very few birds were recorded during nocturnal surveys. These included Marbled *Podargus ocellatus* and Papuan Frogmouth *P. papuensis*, and the Papuan Boobook *Ninox theomacha*.

The total of 142 species compares favorably with surveys elsewhere in New Guinea. Beehler and Mack (1998) obtained similar or smaller totals for surveys of the same duration in the Lakekamu Basin (PNG). Archbold collected

1000 skins of about 150 bird species at Bernhard camp in six weeks of collecting. The full day count on 12 September produced a list of 54 species, and a trip to Buare Lagoon that encompassed a wide range of habitats recorded 61 species. These results, and the failure of our species accumulation curve to asymptote, suggest that local (alpha) species diversity in the area is quite high.

The Bird Trade

We briefly surveyed the trade in birds for pets and souvenirs (e.g. stuffed specimens). Prices for Lesser Birds of Paradise, Twelve-wired Birds of Paradise, and Cockatoos ranged between Rp. 30,000 and 60,000 (ca USD $3–6). Cassowaries, crowned pigeons, and hornbills are hunted predominantly for the local market as a protein supplement, and are among the first species to disappear around settlements in New Guinea. Surprisingly, these species were relatively common within one to two hours walking distance from Dabra village. Hunting does occur in the area, and a cassowary was butchered by local hunters near the Furu Camp during our survey. The relatively insignificant impact of subsistence hunting on local avifauna is probably due to the reliance of local people on freshwater fish for protein, and to the low human population density. Mamberamo Hulu district comprises 14 villages, numbering just 7,800 people in an area of 11,095 km² (Mamberamo Hulu district office, unpublished data). However, hunting to supply the pet and souvenir industries may become a significant threat in the future because freshwater crocodiles (the main source of income for most people in the Mamberamo Basin) are becoming rare due to over harvesting. Habitat destruction by commercial logging and the imminent construction of the Mamberamo Dam are threats of a very different scale, and have the potential to cause far more damage to the environment than any hunting practices used by local communities.

Ethno-ornithology

Our guides, especially the older ones, are knowledgeable naturalists, and they detected several significant species that were overlooked by the scientists during this survey. The Dasigo and Airo bird names are often very similar, indicating the close relationship between the two languages. In contrast the Kaowerawej names are very different. Our guides had different names for most species, although some unrelated species shared the same names. For example *suwi*, used for gerygones, is also used for flowerpeckers, sunbirds, a cuckoo, and a triller. This overlap is not based on general morphological resemblance. Clearly more linguistic research is needed.

CONSERVATION RECOMMENDATIONS

The extensive swamps and swamp forests of the Mamberamo Basin harbor bird species not known elsewhere. Some of these were not recorded during the present survey (notably Pale-billed Sicklebill and Brass's Friarbird), and more surveys are needed to document their occurrence in the extensive swamps north of the survey area. Plans to develop the area (dam construction, large-scale logging) should be closely monitored, and opposed or amended if they seriously threaten the survival of Mamberamo's unique birds and other wildlife.

LITERATURE CITED

Archbold, R., A.L. Rand, and L.J. Brass. 1942. Results of the Archbold Expeditions No. 41. Summary of the 1938–1939 New Guinea Expedition. Bull. Am. Mus. Nat. Hist. 79: 199–288.

Beehler, B.M., and A.L. Mack. 1998. Constraints to characterizing spatial heterogeneity in a lowland forest avifauna in New Guinea. *In:* Adams, N.J. and R.H.Slotow, (eds.). Proc. 22 Int. Ornitholog. Congr., Durban: BirdLife South Africa. Pp. 2569–2579.

Conservation International. 1999. The Irian Jaya Biodiversity Conservation Priority-setting Workshop. Biak, 7–12 January 1997. Washington, DC: Conservation International.

Diamond, J.M. 1980. Proposal for a reserve in the Mamberamo region, Irian Jaya. Special report No. 3. Indonesia: World Wildlife Fund.

Diamond, J.M., and K.D. Bishop. 1999. Ethno-ornithology of the Ketengban People, Indonesian New Guinea. *In:* Medin, D.L. and S. Atran (eds.). Folkbiology. Cambridge, Massachusetts: Massachusetts Institute of Technology Press. Pp. 17–45.

Ginn, H.B., and D.S. Melville. 1983. Moult in birds. British Trust for Ornithology Guide 19.

Hartert, E. 1932. Liste der Vögel aus Neuguinea im Buitenzorger Museum in Java. Nova Guinea 15 (5): 435–484.

Mack, A.L., W. Widodo, and Boeadi. 2000. Noteworthy bird species observed on the RAP survey in the Wapoga area, Irian Jaya, Indonesia. *In:* A.L. Mack and L.E. Alonso (eds.). A biological assessment of the Wapoga River area of northwestern Irian Jaya, Indonesia. RAP Bulletin of Biological Assessment 14. Washington, DC: Conservation International. Pp. 127–128.

Stattersfield, A.J., M.J. Crosby, A.J. Long, and D.C. Wege. 1998. Endemic Bird Areas of the World. Priorities for Biodiversity Conservation. Birdlife Conservation Series No. 7.

Stattersfield, A.J. and D.R. Capper. 2000. Threatened birds of the world. Cambridge: BirdLife International.

van Bemmel, A.C.V. 1947. Two small collections of New Guinea Birds. II. The birds of Mamberamo river. Treubia 19: 17–45.

van Heurn, W.C. 1921a. De Expeditie 1920–1921 naar Nieuw-Guinea. Jaarb. Club van Ned. Vogel. (Feestn): 11–28.

van Heurn, W.C. 1921b. Over de vogels van het Mamberamo-gebied (Noord-en Centraal Nederlandsch Nieuw-Guinea). Jaarb. Club van Ned. Vogel. (Feestn): 29–64.

van Heurn, W.C. 1921c. De strooperij in dienst der Ornithologie. Jaarb. Club van Ned. Vogel. (Feestn): 65–69.

Chapter 12

Small mammals of the Dabra area, Mamberamo River Basin, Papua, Indonesia

Rose Singadan and Freddy Patiselanno

CHAPTER SUMMARY

- Sixty-nine mammals, including four species of bats, two species of rodents, and one species of marsupial, were trapped during the survey.

- Documentation of the feather-tailed possum *Distoechurus pennatus* at Furu River fills a wide gap in this species' known distribution and is one of few records for Papua.

- The White-bellied Melomys (*Melomys leucogaster*) was recorded from north of the central cordillera for the first time.

RINGKASAN BAB – MAMALIA DABRA

- Dalam survei ini berhasil ditangkap 69 ekor mamalia, termasuk empat spesies kelelawar, dua spesies rodent/hewan pengerat, dan satu spesies marsupial.

- Dokumentasi Posum Berekor Bulu *Distoechurus pennatus* di Sungai Furu mengisi celah lebar dari distribusi yang telah diketahui untuk spesies tersebut dan hal ini merupakan salah satu dari beberapa rekor untuk Papua.

- Melomis perut putih (*Melomys leucogaster*) tercatat untuk pertama kalinya di sebelah utara dari Cordillera tengah.

INTRODUCTION

Papua encompasses a wide variety of ecosystems, generated in part by a diverse and dramatic topography. The altitudinal range extends from sea level to nearly 5000 m asl, providing a diversity of habitats that have promoted the evolution of many unique and endemic animal groups (Petozc, 1987). Recent surveys have improved our knowledge about the distribution and diversity of mammals in New Guinea (e.g. Flannery and Seri, 1990; Kitchener et al., 1998). The publications of Bonaccorso (1998) and Flannery (1995) have summarized existing information about all, or significant components of, New Guinea's mammal fauna and provide a firm basis for future research. Despite these advances there are still many areas in Papua that have not been adequately sampled.

One such area is the Mamberamo River Basin, a broad expanse of lowland rainforest on the northern side of Papua's central mountains. The aim of this project was to document the diversity of small mammals at two sites near the town of Dabra, and to compare this diversity with that reported from similar lowland habitats elsewhere in New Guinea.

METHODS

The survey was conducted for seven nights between 1–15 September 2000. Small mammals were trapped in primary forest around the Furu River site (four nights), and Tiri River site (three nights) using 71 Elliot traps and five mist nets. Thirty-six Elliot traps were installed in a 50 x 50 m grid, and 35 were set at 2 m intervals along a 70 m transect. A variety of baits were used, including bananas, fish, and biscuits. Traps were opened in the evening (4:00–5:00 p.m.) and were checked and closed the next morning between 8:00–9:00 a.m. Mist nets were erected around the camps and on ridges or across creeks, because these habitats are frequently used as flight corridors by bats. Mist nets were opened at 5:00 pm, and checked again at 9:00 p.m., 1:00 a.m., and 5:00 a.m. during each night.

Each captured mammal was identified using Flannery (1995) and Menzies and Dennis (1979). Several specimens retained as vouchers are deposited in the Museum Zoologie Bogor.

RESULTS AND DISCUSSION

Sixty-nine mammals representing seven species were trapped during the survey, and an additional two species of megachiropteran bats were seen but not collected. In total four species of bats, two species of rodents, and one species of marsupial were positively identified (Table 12.1). Fauna at the two sites was similar; the same four bat species were collected at each site, but no rodents or marsupials were collected at the Tiri River site. All of the bats are megachi-

ropteran fruit bats in the family Pteropodidae. In Papua New Guinea (PNG), approximately 60% of bats are insect-eating microchiropterans (Bonnaccorso, 1998). Given a similar proportion in Papua (Flannery, 1995), it appears that our trapping technique was relatively unsuccessful for insectivorous bats.

One female *Syconycteris australis* at Furu was carrying an embryo, and one *S. australis* and one *Paranyctimene raptor* were lactating. Eight *Nyctimene draconilla* were also carrying young embryos.

The waterside rat (*Parahydromis asper*) has a wide distribution across New Guinea, but collection of the feather-tailed possum *Distoechurus pennatus* at Furu Camp fills a wide gap in its known distribution and is one of few records for this species in Papua (Flannery, 1995). Both *D. pennatus* and *P. asper* are capable of surviving in disturbed forest habitats, and their occurrence at Furu River and the absence of other primary-forest specialists probably reflects a history of forest damage through resource extraction by local communities in this area. The White-bellied Melomys (*Melomys leucogaster*) was previously known only from southern New Guinea (Flannery, 1995), and this is the first record from north of the central cordillera.

Survey effort at the Dabra sites was too limited to produce a comprehensive inventory. However documentation of two significant distributional records indicate that additional surveys are likely to produce more exciting discoveries in this poorly known area.

Table 12.1. Mammals recorded from the Dabra area, Papua, Indonesia.

Species	Common Name	Furu Male/Female	Tiri Male/Female	Comments
Chiroptera	**Bats**			
Syconycteris australis		10 /4	8 / 6	
Paranyctimene raptor		2 /1	2 / 2	
Nyctimene draconilla		5 /9	1 / 0	
Nyctimene albiventer		7 /6	3 / 0	
Marsupialia	**Marsupials**			
Distoechurus pennatus	Feather-tailed possum	0/1	0/0	carrying a young embryo
Rodentia	**Rodents**			
Parahydromis asper	Waterside rat	1/0	0/0	
Melomys leucogaster	White-bellied melomys	1/0	0/0	
Total		**26/21**	**14/8**	

LITERATURE CITED

Bonaccorso, F.J. 1998. Bats of Papua New Guinea. Washington, DC: Conservation International.

Flannery, T.F. 1995. Mammals of New Guinea. Chatswood: Reed Books.

Flannery, T.F. and L. Seri. 1990. The mammals of southern West Sepik Province, Papua New Guinea: their distribution, abundance, human use and zoogeography. Rec. Aust. Mus. 42: 173 208.

Kitchener, D.J., Boeadi and M. Smaga. 1998. The mammals of the PT Freeport Indonesia abstract of work mining and project area, Irian Jaya Indonesia. Biodiversity study series. Vol. 6. Jakarta: PT Freeport.

Menzies, J and E. Dennis. 1979. Handbook of New Guinea rodents. Papua New Guinea: Wau Ecology Institute.

Petocz, R. 1987. Mamalia Darat Irian Jaya. PT. Jakarta: Gramedia Pustaka Utama.

Gazetteer of Focal Sites, Yongsu and Mamberamo

YEMANG CAMP, YONGSU DOSOYO, NORTHERN EDGE OF CYCLOPS MOUNTAINS

02°25.994'S, 140°29.147'E. Elevation 0 to ~ 80 m a.s.l.

The site of Yemang Training Camp. Lowland primary forest and patches of secondary forest and garden re-growth. Evidence of timber extraction throughout most of the forest. This area of forest is referred to as Yongsu Dosoyo (the name of the nearby village) in this report, but does not indicate that Jari forest is located further from the village (Map 1). Forest adjacent to this site was surveyed intensively for most taxa, except vegetation.

JARI FOREST, NORTHERN EDGE OF CYCLOPS MOUNTAINS

02°26.295'S, 140°30.872'E. Altitude 100 m a.s.l.

Predominantly primary lowland rainforest, on a low ridge across the bay from Yemang Camp. A small trail runs the length of the ridge but there is little sign of habitat modification. Vegetation plots and flora surveys were undertaken primarily in this forest, although some surveys were also undertaken adjacent to Yemang Camp.

FURU RIVER, MAMBERAMO BASIN

03°17'04"S, 138°38'10"E. Altitude 90 m a.s.l.

Furu Camp was 3.0 km southeast of Dabra, on the banks of the Furu River, a small tributary of the Idenburg River. Forest around this camp has been extensively disturbed by harvesting of fruit and timber by the Dabra community. The site provided access to swampy lowland forest, secondary forest, and hill forest at the base of the central cordillera.

TIRI RIVER, MAMBERAMO BASIN

03°17'30"S, 138°34'53"E. Altitude 80 m a.s.l.

Tiri Camp was 4.5 km southwest of Dabra on the banks of the Tiri River, a large stream in the catchment of the Doorman River, a major tributary of the Mamberamo River. Forest around the camp is lowland primary rainforest, and evidence of human disturbance is scarce.

DABRA VILLAGE AND VICINITY, MAMBERAMO BASIN

03°16'11.8"S, 138°36'52.7"E Altitude 80 m a.s.l.

Dabra is the major population centre of the region and is situated on the southern bank of the Idenburg River. The immediate surroundings of the village have been extensively modified for gardens. Brief collections of some taxa were made in and around the village, and on the south side of a heavily forested ridge (~ 200 m a.s.l.) south of the village.

Appendices

Appendix 1 96
Seedlings and grasses (≤ 50 cm in height) sampled in
10 sub-plots of 0.5 x 2 m in each of five 0.1 ha plots in Jari
forest, Yongsu Dosoyo, Papua, Indonesia

*Yance de Fretes, Conny Kameubun, Ismail A. Rachman,
Julius D. Nugroho, Elisa Wally, Herman Remetwa,
Marthen Kabiay, Ketut G. Suartana, and Basa T.
Rumahorbo*

Appendix 2 98
Saplings (1.0–4.9 cm dbh) in 2 sub-plots of 2 x 5 m in each
of five 0.1 ha plots in Jari forest, Yongsu Dosoyo, Papua,
Indonesia

*Yance de Fretes, Conny Kameubun, Ismail A. Rachman,
Julius D. Nugroho, Elisa Wally, Herman Remetwa,
Marthen Kabiay, Ketut G. Suartana, and Basa T.
Rumahorbo*

Appendix 3 100
Poles (trees 5.0–9.9 cm dbh) sampled in 5 sub-plots of
5 x 20 m in each of five 0.1 ha plots in Jari forest, Yongsu
Dosoyo, Papua, Indonesia

*Yance de Fretes, Conny Kameubun, Ismail A. Rachman,
Julius D. Nugroho, Elisa Wally, Herman Remetwa,
Marthen Kabiay, Ketut G. Suartana, and Basa T.
Rumahorbo*

Appendix 4 101
Trees (dbh ≥ 10 cm) sampled in 5 sub-plots of 20 x 50 m
and 5 sub-plots of 5 x 20 m in each of five 0.1 ha plots in
Jari forest, Yongsu Dosoyo, Papua, Indonesia

*Yance de Fretes, Conny Kameubun, Ismail A. Rachman,
Julius D. Nugroho, Elisa Wally, Herman Remetwa,
Marthen Kabiay, Ketut G. Suartana, and Basa T.
Rumahorbo*

Appendix 5 104
Beach vegetation observed around Yemang (Yongsu)
Training Camp, Papua, Indonesia

*Yance de Fretes, Conny Kameubun, Ismail A. Rachman,
Julius D. Nugroho, Elisa Wally, Herman Remetwa,
Marthen Kabiay, Ketut G. Suartana, and Basa T.
Rumahorbo*

Appendix 6 105
Vegetation recorded from secondary forests around
Yemang Training Camp, Yongsu Dosoyo, Papua, Indonesia

*Yance de Fretes, Conny Kameubun, Ismail A. Rachman,
Julius D. Nugroho, Elisa Wally, Herman Remetwa,
Marthen Kabiay, Ketut G. Suartana, and Basa T.
Rumahorbo*

Appendix 7 106
Vegetation recorded outside of plots during surveys
of forests at Jari and Yemang, Yongsu area, Papua,
Indonesia

*Yance de Fretes, Conny Kameubun, Ismail A. Rachman,
Julius D. Nugroho, Elisa Wally, Herman Remetwa,
Marthen Kabiay, Ketut G. Suartana, and Basa T.
Rumahorbo*

Appendix 8 109
Grasses, seedlings, and herbaceous plants (≤ 50 cm
in height) in 50 subplots (0.5 x 2 m) in five 20 x 50 m
Whittaker Plots at Furu River, Papua, Indonesia

Yance de Fretes, Ismail A. Rachman, and Elisa Wally

Appendix 9 111
List of all trees (≥ 1 cm dbh) sampled in 5 Whitaker plots
(Furu and Jari) and 2 transects (Tiri) during 2000 RAP
surveys, Papua, Indonesia

Yance de Fretes, Ismail A. Rachman, and Elisa Wally

Appendix 10 119
Botanical specimens collected outside of plots during
Papua (Indonesia) RAP surveys (2000) and identified by
Ismail A. Rachman

Yance de Fretes, Ismail A. Rachman, and Elisa Wally

Appendix 11   127
Plant species grown in gardens and used by the Dabra
community (Mamberamo Basin), Papua, Indonesia

Yance de Fretes, Ismail A. Rachman, and Elisa Wally

Appendix 12 129
Sampling stations for aquatic insect surveys in the Dabra
area, Papua, Indonesia

Dan A. Polhemus

Appendix 13 130
Aquatic insects captured at nine sampling stations in the
Dabra area, Papua, Indonesia

Dan A. Polhemus

Appendix 14 133
Annotated checklist of aquatic insects collected during
the Dabra RAP survey, Papua, Indonesia

Dan A. Polhemus

Appendix 15 137
Aquatic insects collected in the Cyclops Mountains,
Papua, Indonesia

Dan A. Polhemus

Appendix 16 138
List of butterflies collected around Yongsu Dosoyo,
Papua, Indonesia

*Edy Rosariyanto, Henk van Mastrigt, Henry Silka Innah,
and Hugo Yoteni*

Appendix 17 140
List of butterflies recorded at Furu and Tiri Rivers,
Mamberamo Basin, Papua, Indonesia

Henk van Mastrigt and Edy M. Rosariyanto

Appendix 18 142
Diversity of moths collected in the Dabra area, Papua,
Indonesia

Henk van Mastrigt and Edy M. Rosariyanto

Appendix 19 144
Summary of fish collection/observation sites in the Yongs
u and Dabra areas, Papua, Indonesia

Gerald R. Allen

Appendix 20 146
Summary of freshwater fishes collected during the
Yongsu training course, Papua, Indonesia

Gerald R. Allen

Appendix 21 148
Annotated checklist of fishes of the Yongsu area, Papua,
Indonesia

Gerald R. Allen

Appendix 22 151
List of shallow coral reef fishes of Yongsu Bay, Papua,
Indonesia

Gerald R. Allen

Appendix 23 155
Summary of fishes collected on the RAP survey in the
Mamberamo River drainage, Papua, Indonesia

Gerald R. Allen

Appendix 24 157
Annotated checklist of fishes recorded from the
Mamberamo River system, Papua, Indonesia

Gerald R. Allen

Appendix 25 160
List of fish species recorded to date from the
Mamberamo, Sepik, and Ramu Rivers of northern New
Guinea

Gerald R. Allen

Appendix 26 162
Distribution of amphibians and reptiles in the Yongsu area
(0–400 m asl), Papua, Indonesia

*Stephen Richards, Djoko Iskandar, Burhan Tjaturadi, and
Aditya Krishar*

Appendix 27 164
Frogs and reptiles recorded from three sites in the Dabra
area, Papua, Indonesia

Stephen Richards, Djoko Iskandar, and Burhan Tjaturadi

Appendix 28 166
Annotated list of noteworthy frogs and reptiles recorded
from three sites in the Dabra area, Papua, Indonesia

Stephen Richards, Djoko Iskandar, and Burhan Tjaturadi

Appendix 29 168
Birds recorded at Yongsu, Papua, Indonesia

*Pujo Setio, Paul Johan Kawatu, Irba U. Nugroho, David
Kalo, Daud Womsiwor, and Bruce M. Beehler*

Appendix 30 .... **71**

Forest trees at two bird census points, Yongsu, Papua, Indonesia

Pujo Setio, Paul Johan Kawatu, Irba U. Nugroho, David Kalo, Daud Womsiwor, and Bruce M. Beehler

Appendix 31 .... **172**

Birds recorded from the Mamberamo/Idenburg river basins, Papua, Indonesia

Bas van Balen, Suer Suryadi, and David Kalo

Appendix 32 .... **179**

Annotated list of noteworthy birds known from or expected to occur in the Dabra area

Bas van Balen, Suer Suryadi, and David Kalo

Appendix 1

Seedlings and grasses (≤ 50 cm in height) sampled in 10 sub-plots of 0.5 x 2 m in each of five 0.1 ha plots in Jari forest, Yongsu Dosoyo, Papua, Indonesia

Yance de Fretes, Conny Kameubun, Ismail A. Rachman, Julius D. Nugroho, Elisa Wally, Herman Remetwa, Marthen Kabiay, Ketut G. Suartana, and Basa T. Rumahorbo

Values indicate number of individuals recorded.

| Family | Species | Plot Number | | | | | |
		1	2	3	4	5	Total
Anacardiaceae	*Campnosperma auriculata*				1		1
Annonaceae	*Polyalthia glauca*				4		4
Araceae	*Alocasia* sp.	1					1
Araceae	*Rhaphidophora* sp.	1					1
Arecaceae	*Caryota rumphiana*		1				1
Arecaceae	*Licuala* sp.		1		5	4	10
Arecaceae	*Pinanga* sp.					2	2
Arecaceae	*Rhopaloblaste* sp.	8	4	16	3	7	38
Burseraceae	*Canarium* sp.	6	75	49	40	520	690
Burseraceae	*Haplolobus* sp.	3	4	1			8
Clusiaceae	*Calophyllum* sp.	3		1	1	1	6
Clusiaceae	*Garcinia* sp.		2	3	6		11
Commelinaceae	*Commelina nudiflora*					1	1
Ebenaceae	*Diospyros discolor*	3	31	17	9	5	65
Elaeocarpaceae	*Elaeocarpus* sp.	5	13	13	23	15	69
Euphorbiaceae	*Pimelodendron amboinicum*				2		2
Fabaceae	*Adenanthera* sp.	1					1
Fabaceae	*Cynometra ramiflora*	4	3	2	3	1	13
Fabaceae	*Intsia bijuga*				1		1
Gleicheniaceae	*Dicranopteris liniaris*	1	1				2
Gnetaceae	*Gnetum gnemon*	4					4
Icacinaceae	*Gomphandra* sp.	1		2	5	1	9
Icacinaceae	*Gonocaryum* sp.	3	1	3	1	10	18
Lauraceae	*Actinodaphne* sp.	2		1			3
Lauraceae	*Cryptocarya* sp.				3	1	4
Lauraceae	*Litsea* sp.	34	3	6	2	1	46

continued

Family	Species	Plot Number					
		1	2	3	4	5	Total
Liliaceae	*Dianela ensifolia*				2		2
Loganiaceae	*Fagraea racemosa*				1		1
Meliaceae	*Dysoxylum* sp.			1			1
Minispermaceae	*Minisperma* sp.			1			1
Moraceae	*Ficus* sp.				1	4	5
Myristicaceae	*Horsfieldia* sp.		1	2	5	3	11
Myristicaceae	*Myristica* sp.	1	2	3	1		7
Myrtaceae	*Syzygium* sp.	22	23	9	12	18	84
Pandanaceae	*Freycinetia* sp.	8	2		5	2	17
Pandanaceae	*Pandanus* sp.	2			1		3
Podocarpaceae	*Podocarpus blulmei*		2	3	5	4	14
Polypodiaceae	*Dryopteris* sp.				2		2
Polypodiaceae	*Nephrolepsis* sp.				2		2
Ranunculaceae	*Clematis* sp.		1				1
Rubiaceae	*Mastixiodendron* sp.	1		19	1	5	26
Sapindaceae	*Jagera* sp.				1		1
Sapindaceae	*Pometia pinnata*	2	2	3	7	2	16
Sapotaceae	*Manilkara* sp.		1		6	3	10
Sapotaceae	*Palaquium* sp.	5	1	16	15	2	39
Selaginellaceae	*Selaginella* sp.				1		1
Sterculiaceae	*Sterculia* sp.	1	1				2
Ulmaceae	*Celtis* sp.					13	13
Verbenaceae	*Teijsmanniodendron* sp.				2		2
Zingiberaceae	*Amomum* sp.	2					2
Total	**Species**	25	20	23	34	23	50
	Individuals	124	173	173	179	625	1274

Appendix 2

Saplings (1.0–4.9 cm dbh) in 2 sub-plots of 2 x 5 m in each of five 0.1 ha plots in Jari forest, Yongsu Dosoyo, Papua, Indonesia

Yance de Fretes, Conny Kameubun, Ismail A. Rachman, Julius D. Nugroho, Elisa Wally, Herman Remetwa, Marthen Kabiay, Ketut G. Suartana, and Basa T. Rumahorbo

Values indicate number of individuals recorded.

| Family | Species | Plot Number | | | | | |
		1	2	3	4	5	Total
Apocynaceae	*Cerbera floribunda*				1		1
Arecaceae	*Licuala* sp.			1			1
Arecaceae	*Pinanga* sp.		1				1
Arecaceae	*Rhopaloblaste* sp.	1					1
Burseraceae	*Canarium* sp.		4	3	2	5	14
Clusiaceae	*Calophyllum* sp.					2	2
Clusiaceae	*Garcinia* sp.		2	2	1	1	6
Cunoniaceae	*Schizomeria* sp.				1		1
Dilleniaceae	*Dillenia* sp.	2					2
Ebenaceae	*Diospyros* sp.		1				1
Elaeocarpaceae	*Elaeocarpus* sp.				1		1
Euphorbiaceae	*Glochidion* sp.	1			1		2
Fabaceae	*Adenanthera* sp.					1	1
Flacourtiaceae	*Homalium* sp.					1	1
Gnetaceae	*Gnetum gnemon*			2			2
Icacinaceae	*Gomphandra* sp.			3	2	1	6
Icacinaceae	*Gonocaryum* sp.				1	1	2
Lauraceae	*Cinnamomum* sp.	1					1
Lauraceae	*Cryptocarya* sp.				2		2
Lecythidaceae	*Barringtonia* sp.					1	1
Melastomataceae	*Medinilla* sp.			1			1
Meliaceae	*Aglaia* sp.	1	1	1			3
Meliaceae	*Dysoxylum* sp.			1	1	2	4

continued

Family	Species	Plot Number					
		1	2	3	4	5	Total
Myristicaceae	*Horsfieldia* sp.	1		1			2
Myristicaceae	*Myristica* sp.		1	2	3		6
Myrtaceae	*Decaspermum* sp.				1		1
Myrtaceae	*Syzygium* sp.	2	1	1	1		5
Rubiaceae	*Mastixiodendron* sp.			1	1		2
Sapindaceae	*Pometia* sp.		1	2	1		4
Sapotaceae	*Manilkara* sp.					1	1
Sapotaceae	*Palaquium* sp.				1	2	3
Ulmaceae	*Celtis* sp.		1				1
Total	**Species**	7	9	13	16	11	32
	Individuals	9	13	21	21	18	82

Appendix 3

Poles (trees 5.0–9.9 cm dbh) sampled in 5 sub-plots of 5 x 20 m in each of five 0.1 ha plots in Jari forest, Yongsu Dosoyo, Papua, Indonesia

Yance de Fretes, Conny Kameubun, Ismail A. Rachman, Julius D. Nugroho, Elisa Wally, Herman Remetwa, Marthen Kabiay, Ketut G. Suartana, and Basa T. Rumahorbo

Family	Species	Plot Number 1	2	3	4	5	Total
Anacardiaceae	*Mangifera* sp.					1	1
Arecaceae	*Licuala* sp.				1		1
Arecaceae	*Pinanga* sp.	1					1
Burseraceae	*Canarium maluense*		1				1
Burseraceae	*Canarium* sp.	1		1	1	1	4
Clusiaceae	*Calophyllum* sp.			1			1
Clusiaceae	*Garcinia celebica*					1	1
Clusiaceae	*Garcinia* sp.	1					1
Euphorbiaceae	*Pimeleodendron amboinicum*					1	1
Gnetaceae	*Gnetum gnemon*				1		1
Icacinaceae	*Gomphandra montana*		1				1
Icacinaceae	*Gonocaryum* sp.			1	1		2
Lecythidaceae	*Barringtonia racemosa*		1				1
Lecythidaceae	*Barringtonia* sp.	1					1
Melastomataceae	*Medinila* sp.			1			1
Meliaceae	*Dysoxylum* sp.	1					1
Myristicaceae	*Gymnacranthera paniculata*					1	1
Myristicaceae	*Myristica lancifolia*		1		1		2
Myristicaceae	*Myristica* sp.			1		1	2
Myrtaceae	*Syzygium* sp.				2	1	3
Proteaceae	*Helicia* sp.		1				1
Rosaceae/Chrysobalanaceae	*Prunus arborea*				1		1
Rubiaceae	*Timonius* sp.				1		1
Rutaceae	*Tetractomia* sp.				1		1
Sapindaceae	*Pometia pinnata*		1				1
Sterculiaceae	*Sterculia* sp.	1					1
Verbenaceae	*Teijsmanniodendron ahernianum*		1				1
Total	**Species**	6	7	5	9	7	27
	Individuals	6	7	5	10	7	35

Appendix 4

Trees (dbh ≥ 10 cm) sampled in 5 sub-plots of 20 x 50 m and 5 sub-plots of 5 x 20 m in each of five 0.1 ha plots in Jari forest, Yongsu Dosoyo, Papua, Indonesia

Yance de Fretes, Conny Kameubun, Ismail A. Rachman, Julius D. Nugroho, Elisa Wally, Herman Remetwa, Marthen Kabiay, Ketut G. Suartana, and Basa T. Rumahorbo

Values indicate number of individuals recorded.

| Family | Species | Plot | | | | | |
		1	2	3	4	5	Total
Anacardiaceae	*Campnosperma auriculata*		1	1	3		5
Anacardiaceae	*Campnosperma brevipetiolata*	1					1
Annonaceae	*Mezzetia* sp.				1		1
Annonaceae	*Polyalthia glauca*		1		1	1	3
Annonaceae	*Xylopia* sp.					2	2
Arecaceae	*Licuala* sp.		1	2	1		4
Arecaceae	*Rhopaloblaste* sp.					1	1
Burseraceae	*Canarium maluense*	1	2	3			6
Burseraceae	*Canarium oleosum*				2	2	4
Burseraceae	*Canarium* sp.			1	2	1	4
Burseraceae	*Santiria* sp.		2		1	1	4
Clusiaceae	*Calophyllum soulatri*	1	2				3
Clusiaceae	*Calophyllum* sp. 1			1		1	2
Clusiaceae	*Calophyllum* sp. 2				1		1
Clusiaceae	*Garcinia celebica*	1	1	3	1		6
Clusiaceae	*Garcinia* sp. 1	1	2				3
Clusiaceae	*Garcinia* sp. 2		1				1
Clusiaceae	*Garcinia* sp. 3		1				1
Clusiaceae	*Garcinia* sp. 4			1			1
Cunoniaceae	*Schizomeria* sp.				1		1
Cunoniaceae	unidentified		1			1	2
Dilleniaceae	*Dillenia quercifolia*		1				1
Dipterocarpaceae	*Hopea novoguenensis*	2		1		4	7
Ebenaceae	*Diospyros discolor*	1					1
Ebenaceae	*Diospyros* sp. 1	1				6	7
Euphorbiaceae	*Drypetes longifolia*		1				1
Euphorbiaceae	*Drypetes* sp. 1		2		1		3
Euphorbiaceae	*Glochidion* sp.	1	1				2
Euphorbiaceae	*Neoscortechinia*		1		1		2

continued

Family	Species	Plot 1	2	3	4	5	Total
Euphorbiaceae	*Pimelodendron amboinicum*	1			2		3
Euphorbiaceae	unidentified		1				1
Fabaceae	*Cynometra* sp.	1	3	1	2		7
Gnetaceae	*Gnetum* sp.	3	2	2		2	9
Icacinaceae	*Gomphandra montana*		1		1		2
Icacinaceae	*Gonocaryum* sp. 1		1				1
Icacinaceae	*Gonocaryum* sp. 2		1				1
Icacinaceae	*Gonocaryum* sp. 3		1		2		3
Icacinaceae	*Platea exelsa*	1			1	1	3
Icacinaceae	*Stemonurus* sp.		2				2
Lauraceae	*Cryptocarya* sp.		2			2	4
Lauraceae	*Endiandra* sp.	1					1
Lauraceae	*Litsea* sp.	2		1	1		4
Lecythidaceae	*Barringtonia racemosa*	1			1	1	3
Lecythidaceae	*Barringtonia* sp. 1			2	2		4
Loganiaceae	*Fagraea racemosa*		2				2
Melastomataceae	*Astronia* sp.		1	1			2
Meliaceae	*Chisocheton* sp.	1	1				2
Meliaceae	*Dysoxylum* sp.	2				1	3
Moraceae	*Prainea papuana*	1	3	1		1	6
Myristicaceae	*Gymnacranthera*	2		9	12	4	27
Myristicaceae	*Horsfieldia* sp. 1			1		3	4
Myristicaceae	*Horsfieldia* sp. 2			2			2
Myristicaceae	*Myristica lancifolia*		1	2			3
Myristicaceae	*Myristica* sp.		1	4		4	9
Myrtaceae	*Syzygium* sp. 1	1					1
Myrtaceae	*Syzygium* sp. 2	1					1
Myrtaceae	*Syzygium* sp. 3	1					1
Myrtaceae	*Syzygium* sp. 4		1		1		2
Myrtaceae	unidentified		1				1
Oleaceae	*Cionanthus* sp.	1			1		2
Pandanaceae	*Pandanus* sp.			1		2	3
Podoccarpaceae	*Podocarpus blumei*	1	1				2
Proteaceae	*Helicia* sp.	1		1			2
Rosaceae	*Parinarium nonda*			1	1		2
Rosaceae	*Parastemon* sp.			1			1
Rosaceae	*Parastemon urophyllus*			1	1		2
Rosaceae	*Parinari* sp.		1	2	1		4
Rosaceae	*Prunus* sp.		2	3			5
Rosaceae	unidentified	1		2		1	4
Rubiaceae	*Adina multifolia*		1	4	1	1	7
Rubiaceae	*Mastixiodendron pachyclados*		1		2	2	5

continued

| Family | Species | Plot | | | | | |
		1	2	3	4	5	Total
Rubiaceae	*Timonius* sp.	1	4			1	6
Rubiaceae	unidentified				5		5
Rutaceae	*Tetractomia* sp.	1	1	1			3
Sapindaceae	*Pometia pinnata*	1					1
Sapindaceae	*Pometia* sp.				1		1
Sapindaceae	unidentified	1	1		1		3
Sapotaceae	*Chrysophyllum fasciculata*			1		1	2
Sapotaceae	*Manilkara fasciculata*			1			1
Sapotaceae	*Palaquium* sp. 1	2					2
Sapotaceae	*Palaquium* sp. 2	1					1
Sapotaceae	*Planchonela* sp.	1					1
Sterculiaceae	*Sterculia* sp.	1	1				2
Theaceae	*Gordonia* sp.	2	1	3	1	1	5
Theaceae	*Ternstroemia* sp.		1		1		2
Ulmaceae	*Celtis philippinensis*				2		2
Verbenaceae	*Tejsmanniodendron aherhianum*	1	2	6	2		11
Verbenaceae	*Tejsmanniodendron sogoriense*	1	1	9			11
Total	**Species**	**38**	**46**	**34**	**35**	**26**	**88**
	Individuals	**46**	**64**	**76**	**61**	**48**	**296**

Appendix 5

Beach vegetation observed around Yemang (Yongsu) Training Camp, Papua, Indonesia

Yance de Fretes, Conny Kameubun, Ismail A. Rachman, Julius D. Nugroho, Elisa Wally, Herman Remetwa, Marthen Kabiay, Ketut G. Suartana, and Basa T. Rumahorbo

Family	Species	Growth form
Apocynaceae	*Cerbera floribunda*	Tree
Apocynaceae	*Lepiniopsis ternatensis*	Tree
Arecaceae	*Metroxylon sagu*	Tree palm
Asteraceae	*Wedelia biflora*	Herb
Borraginaceae	*Cordonia subcordata*	Tree
Cassuarinaceae	*Cassuarina equisitifolia*	Tree
Clusiaceae	*Calophyllum inophyllum*	Tree
Combretaceae	*Terminalia catapa*	Tree
Convolvulaceae	*Ipomoea pes-caprae*	Creeping herb
Cycadaceae	*Cycas campestris*	Tree
Fabaceae	*Inocarpus fagiferus*	Tree
Fabaceae	*Pongamia pinnata*	Tree
Flacourtiaceae	*Scolopia chinensis*	Tree
Godeniaceae	*Scaevola frutescens*	Shrub
Hernadiaceae	*Hernandia ovigera*	Tree
Lecythidaceae	*Barringtonia asiatica*	Tree
Lecythidaceae	*Barringtonia racemosa*	Tree
Liliaceae	*Crinum asiaticum*	Herb
Malvaceae	*Hibiscus tiliaceus*	Tree
Meliaceae	*Xylocarpus molluccensis*	Tree
Myrsinaceae	*Ardisia humilis*	Small tree
Myrsinaceae	*Rapanea* sp.	Small tree
Myrtaceae	*Rhodamnia cinerea*	Tree
Myrtaceae	*Tristaniopsis* sp.	Tree
Pandanaceae	*Pandanus tectorius*	Shrub
Rubiaceae	*Mastixiodendron pachylados*	Tree
Sterculiaceae	*Heriteira littoralis*	Tree
Sterculiaceae	*Sterculia macrophylla*	Tree
Sterculiaceae	*Sterculia* sp.	Tree
Verbenaceae	*Premna corymbosa*	Tree

Appendix 6

Vegetation recorded from secondary forests around Yemang Training Camp, Yongsu Dosoyo, Papua, Indonesia

Yance de Fretes, Conny Kameubun, Ismail A. Rachman, Julius D. Nugroho, Elisa Wally, Herman Remetwa, Marthen Kabiay, Ketut G. Suartana, and Basa T. Rumahorbo

Family	Species	Growth form
Anacardiaceae	*Campnosperma auriculata*	Tree
Anacardiaceae	*Dracontomelon dao*	Tree
Anacardiaceae	*Semecarpus cassuvium*	Tree
Annonaceae	*Canangium odoratum*	Tree
Araliaceae	*Boerlagiodendron moluccanum*	Small tree
Araliaceae	*Gastonia* sp.	Tree
Araliaceae	*Schefflera verstegii*	Woody liana
Asteraceae	*Eupatorium odoratum*	Herb
Convolvulaceae	*Merremia peltata*	Liana
Datiscaceae	*Octomeles sumatrana*	Tree
Dilleniaceae	*Tetracera* sp.	Liana
Euphorbiaceae	*Endospermum peltatun*	Tree
Euphorbiaceae	*Macaranga aleuritoides*	Tree
Euphorbiaceae	*Macaranga* sp.	Tree
Fabaceae	*Phanera* sp.	Woody liana
Fabaceae	*Pterocarpus indicus*	Tree
Liliaceae	*Dracaena angustifolia*	Shrub
Loganiaceae	*Fagraea racemosa*	Small tree
Maranthaceae	*Donax cannaeformis*	Herb
Moraceae	*Ficus capiosa*	Tree
Moraceae	*Ficus damaropsis*	Tree
Moraceae	*Ficus glomerata*	Tree
Moraceae	*Ficus septica*	Shrub
Moraceae	*Ficus variegata*	Tree
Myrsinaceae	*Maesa* sp.	Shrub
Piperaceae	*Piper aduncum*	Shrub
Rhamnaceae	*Alphitonia incana*	Tree
Rubiaceae	*Wendlandia glabra*	Herb
Rutaceae	*Euodia elleryana*	Tree
Sapindaceae	*Jagera serrata*	Small tree
Smilaxaceae	*Smilax leucophylla*	Liana
Tiliaceae	*Trichopermum* sp.	Tree
Ulmaceae	*Trema orientalis*	Tree

Family	Species	Growth form
Urticaceae	*Leucosyke capitelata*	Shrub
Verbenaceae	*Geunsia* sp.	Tree
Verbenaceae	*Lantana camara*	Shrub
Vitaceae	*Tetrastigma lanceorarium*	Liana
Maranthaceae	*Halopegia* sp.	Herb

Appendix 7

Vegetation recorded outside of plots during surveys of forests at Jari and Yemang, Yongsu area, Papua, Indonesia

Yance de Fretes, Conny Kameubun, Ismail A. Rachman, Julius D. Nugroho, Elisa Wally, Herman Remetwa, Marthen Kabiay, Ketut G. Suartana, and Basa T. Rumahorbo

Family	Species	Growth form
Anacardiaceae	*Campnosperma auriculata*	Tree
Anacardiaceae	*Campnosperma brevipetiolata*	Tree
Anacardiaceae	*Dracontomelon da'o*	Tree
Anacardiaceae	*Mangifera* sp.	Tree
Anacardiaceae	*Semecarpus cassuvium*	Tree
Annonaceae	*Canangium odoratum*	Tree
Annonaceae	*Mezzetia* sp.	Tree
Annonaceae	*Polyalthia glauca*	Tree
Annonaceae	*Xylopia* sp.	Tree
Apocynaceae	*Cerbera floribunda*	Tree
Apocynaceae	*Lepiniopsis ternatensis*	Tree
Araliaceae	*Boerlagiodendron moluccanum*	Small tree
Araliaceae	*Gastonia* sp.	Tree
Araliaceae	*Schefflera verstegii*	Woody liana
Arecaceae	*Licuala* sp.	Tree
Arecaceae	*Metroxylon sagu*	Tree palm
Arecaceae	*Pinanga* sp.	Tree
Arecaceae	*Rhopaloblaste* sp.	Small tree
Asteraceae	*Eupatorium odoratum*	Herb
Asteraceae	*Wedelia biflora*	Herb
Borraginaceae	*Cordonia subcordata*	Tree
Burseraceae	*Canarium maluense*	Tree
Burseraceae	*Canarium oleosum*	Tree
Burseraceae	*Canarium* sp.	Tree
Burseraceae	*Sanitaria* sp.	Tree
Cassuarinaceae	*Cassuarina equisitifolia*	Tree
Clusiaceae	*Calophyllum inophyllum*	Tree
Clusiaceae	*Calophyllum soulatri*	Tree
Clusiaceae	*Calophyllum* sp.	Small tree
Clusiaceae	*Calophyllum* sp. 1	Tree

Family	Species	Growth form
Clusiaceae	*Calophyllum* sp. 2	Tree
Clusiaceae	*Garcinia celebica*	Tree
Clusiaceae	*Garcinia* sp. 1	Tree
Clusiaceae	*Garcinia* sp. 2	Tree
Clusiaceae	*Garcinia* sp. 3	Tree
Clusiaceae	*Garcinia* sp. 4	Tree
Combretaceae	*Terminalia catapa*	Tree
Convolvulaceae	*Ipomoea pes-caprae*	Creeping herb
Convolvulaceae	*Merremia peltata*	Liana
Cunoniaceae	*Schizomeria* sp.	Small tree
Cycadaceae	*Cycas campestris*	Tree
Datiscaceae	*Octomeles sumatrana*	Tree
Dilleniaceae	*Dillenia quercifolia*	Tree
Dilleniaceae	*Dillenia* sp.	Small tree
Dilleniaceae	*Tetracera* sp.	Liana
Dipterocarpaceae	*Hopea novoguenensis*	Tree
Ebenaceae	*Diospyros discolor*	Tree
Ebenaceae	*Diospyros* sp.	Small tree
Ebenaceae	*Diospyros* sp. 1	Tree
Elaeocarpaceae	*Elaeocarpus* sp.	Small tree
Euphorbiaceae	*Drypetes longifolia*	Tree
Euphorbiaceae	*Drypetes* sp. 1	Tree
Euphorbiaceae	*Endospermum peltatun*	Tree
Euphorbiaceae	*Glochidion* sp.	Small tree
Euphorbiaceae	*Macaranga aleuritoides*	Tree
Euphorbiaceae	*Macaranga* sp.	Tree
Euphorbiaceae	*Neoscortechinia*	Tree
Euphorbiaceae	*Pimelodendron amboinicum*	Tree
Fabaceae	*Adenanthera* sp.	Small tree
Fabaceae	*Cynometra* sp.	Tree
Fabaceae	*Inocarpus fagiferus*	Tree

continued

Family	Species	Growth form
Fabaceae	*Phanera* sp.	Woody liana
Fabaceae	*Pongamia pinnata*	Tree
Fabaceae	*Pterocarpus indicus*	Tree
Flacourtiaceae	*Homalium* sp.	Small tree
Flacourtiaceae	*Scolopia chinensis*	Tree
Gnetaceae	*Gnetum gnemon*	Small tree
Godeniaceae	*Scaevola frutescens*	Shrub
Hernadiaceae	*Hernandia ovigera*	Tree
Icacinaceae	*Gomphandra montana*	Tree
Icacinaceae	*Gomphandra* sp.	Small tree
Icacinaceae	*Gonocaryum* sp.	Tree
Icacinaceae	*Gonocaryum* sp. 1	Tree
Icacinaceae	*Gonocaryum* sp. 2	Tree
Icacinaceae	*Gonocaryum* sp. 3	Tree
Icacinaceae	*Platea exelsa*	Tree
Icacinaceae	*Stemonurus* sp.	Tree
Lauraceae	*Cinnamomum* sp.	Small tree
Lauraceae	*Cryptocarya* sp.	Small tree
Lauraceae	*Endiandra* sp.	Tree
Lauraceae	*Litsea* sp.	Tree
Lecythidaceae	*Barringtonia asiatica*	Tree
Lecythidaceae	*Barringtonia racemosa*	Tree
Lecythidaceae	*Barringtonia* sp.	Small tree
Lecythidaceae	*Barringtonia* sp. 1	Tree
Liliaceae	*Crinum asiaticum*	Herb
Liliaceae	*Dracaena angustifolia*	Shrub
Loganiaceae	*Fagraea racemosa*	Small tree
Malvaceae	*Hibiscus tiliaceus*	Tree
Maranthaceae	*Donax cannaeformis*	Herb
Maranthaceae	*Halopegia* sp.	Herb
Melastomataceae	*Astronia* sp.	Tree
Melastomataceae	*Medinila* sp.	Small tree
Meliaceae	*Aglaia* sp.	Small tree
Meliaceae	*Chisocheton* sp.	Tree
Meliaceae	*Dysoxylum* sp.	Small tree
Meliaceae	*Xylocarpus molluccensis*	Tree
Moraceae	*Ficus capiosa*	Tree
Moraceae	*Ficus dammaropsis*	Tree
Moraceae	*Ficus glomerata*	Tree
Moraceae	*Ficus septica*	Shrub
Moraceae	*Ficus variegata*	Tree
Moraceae	*Prainea papuana*	Tree
Myristicaceae	*Gymnacranthera paniculata*	Tree
Myristicaceae	*Horsfieldia* sp.	Small tree

Family	Species	Growth form
Myristicaceae	*Horsfieldia* sp. 1	Tree
Myristicaceae	*Horsfieldia* sp. 2	Tree
Myristicaceae	*Myristica lancifolia*	Tree
Myristicaceae	*Myristica* sp.	Tree
Myrsinaceae	*Ardisia humilis*	Small tree
Myrsinaceae	*Maesa* sp.	Shrub
Myrsinaceae	*Rapanea* sp.	Small tree
Myrtaceae	*Decaspermum* sp.	Small tree
Myrtaceae	*Rhodamnia cinerea*	Tree
Myrtaceae	*Syzygium* sp.	Tree
Myrtaceae	*Syzygium* sp. 1	Tree
Myrtaceae	*Syzygium* sp. 2	Tree
Myrtaceae	*Syzygium* sp. 3	Tree
Myrtaceae	*Syzygium* sp. 4	Tree
Myrtaceae	*Tristaniopsis* sp.	Tree
Oleaceae	*Cionanthus* sp.	Tree
Pandanaceae	*Pandanus* sp.	Tree
Pandanaceae	*Pandanus tectorius*	Shrub
Piperaceae	*Piper aduncum*	Shrub
Podoccarpaceae	*Podocarpus blumei*	Tree
Proteaceae	*Helicia* sp.	Tree
Proteaceae	*Helicia* sp.	Tree
Rhamnaceae	*Alphitonia incana*	Tree
Rosaceae	*Parinarium nonda*	Tree
Rosaceae	*Parastemon* sp.	Tree
Rosaceae	*Parastemon urophyllus*	Tree
Rosaceae	*Parinari* sp.	Tree
Rosaceae	*Prunus* sp.	Tree
Rosaceae/ Chrysobalanaceae	*Prunus arborea*	Tree
Rubiaceae	*Adina multifolia*	Tree
Rubiaceae	*Mastixiodendron pachylados*	Tree
Rubiaceae	*Mastixiodendron* sp.	Small tree
Rubiaceae	*Timonius* sp.	Tree
Rubiaceae	*Wendlandia glabra*	Herb
Rutaceae	*Eoudia elleryana*	Tree
Rutaceae	*Tetractomia* sp.	Tree
Rutaceae	*Tetractomia* sp.	Tree
Sapindaceae	*Jagera serrata*	Small tree
Sapindaceae	*Pometia pinnata*	Tree
Sapindaceae	*Pometia* sp.	Tree
Sapotaceae	*Chrysophyllum fasciculata*	Tree
Sapotaceae	*Manilkara fasciculata*	Tree
Sapotaceae	*Manilkara* sp.	Small tree

continued

Family	Species	Growth form
Sapotaceae	*Palaquium* sp.	Small tree
Sapotaceae	*Palaquium* sp. 1	Tree
Sapotaceae	*Palaquium* sp. 2	Tree
Sapotaceae	*Planchonela* sp.	Tree
Smilaxaceae	*Smilax leucophylla*	Liana
Sterculiaceae	*Heriteira littoralis*	Tree
Sterculiaceae	*Sterculia macrophylla*	Tree
Sterculiaceae	*Sterculia* sp.	Tree
Theaceae	*Gordonia* sp.	Tree
Theaceae	*Ternstroemia* sp.	Tree
Tiliaceae	*Trichopermum* sp.	Tree
Ulmaceae	*Celtis philippinensis*	Tree
Ulmaceae	*Celtis* sp.	Small tree
Ulmaceae	*Trema orientalis*	Tree
Urticaceae	*Leucosyke capitelata*	Shrub
Verbenaceae	*Geunsia* sp.	Tree
Verbenaceae	*Lantana camara*	Shrub
Verbenaceae	*Premna corymbosa*	Tree
Verbenaceae	*Teijsmanniodendron aherhianum*	Tree
Verbenaceae	*Tejsmanniodendron bogoriense*	Tree
Vitaceae	*Tetrastigma lanceorarium*	Liana

Appendix 8

Grasses, seedlings, and herbaceous plants (≤ 50 cm in height) in 50 subplots (0.5 x 2 m) in five 20 x 50 m Whittaker Plots at Furu River, Papua, Indonesia

Yance de Fretes, Ismail A. Rachman, and Elisa Wally

Number indicates total number of individuals recorded in the five plots.

Family	Species	Number
Adiantaceae	*Adianthum* sp.	10
Annonaceae	*Pseuduvaria* sp. 1	1
Apocynaceae	*Melodinus* sp.	6
Apocynaceae	*Ichnocarpus* sp.	2
Araceae	*Rhaphidophora korthalsii* Schott.	20
Araceae	*Rhaphidophora* sp. 1	12
Araceae	*Rhaphidophora* sp. 2	3
Arecaceae	*Calamus holrungii*	1
Arecaceae	*Calamus* sp. 1	1
Arecaceae	*Calamus* sp. 2	1
Arecaceae	*Calyptrocalyx* sp.	1
Arecaceae	*Gulubia costata* (Becc.) Becc.	3
Arecaceae	*Hydriastele* sp.	2
Arecaceae	*Rhopalobaste* cf. *brassii* H.E. Moore	36
Arecaceae	*Rhopaloblaste* sp. 2	1
Aspleniaceae	*Aspelinidium nidus* L.	1
Aspleniaceae	*Asplenium* cf. *belangeri* (Borry) Kse.	1
Burseraceae	*Canarium oleosum* (Lamk) Engl.	68
Burseraceae	*Canarium* sp.	1
Burseraceae	*Haplolobus floribundus* H.J.L.	5
Burseraceae	*Haplolobus moluccana* H. J. L.	1
Burseraceae	*Haplolobus pachypodum* H.J. Lam	1
Chrysobalanaceae	*Maranthes corymbosa* Bl.	2
Chrysobalanaceae	*Parastemon urophyllus* DC.	1
Chrysobalanaceae	*Parastemon versteghii* M. & P.	2
Chrysobalanaceae	*Prunus arborea* (Bl.) Kalkm.	1
Clusiaceae	*Calophyllum persemile* P. F. Steven	1
Clusiaceae	*Calophyllum* sp. 1	1
Clusiaceae	*Calophyllum* sp. 2	1
Clusiaceae	*Garcinia celebica* Hall. f.	2

Family	Species	Number
Clusiaceae	*Garcinia fruticosa* Lauterb.	1
Clusiaceae	*Garcinia rusteinii* Lauterb.	1
Clusiaceae	*Mamea novoguinensis*	4
Cunnoniaceae	*Ceratopetalum succirubrum* C.T. White	1
Cyatheaceae	*Cyathea* sp.	11
Cyperaceae	*Mapania* sp.	4
Cyperaceae	*Paramapania parvibractea* (Clarke) Uitt.	13
Cyperaceae	*Scleria* sp.	2
Dipteracarpaceae	*Anisoptera thurifera* (Blanco) Blume	11
Dipteracarpaceae	*Hopea novoguinensis* Sloot	7
Dipterocarpaceae	*Shorea* sp.	1
Dipterocarpaceae	*Vatica rassak* (Blanco) Blume	7
Elaeocarpaceae	*Elaeocarpus* sp. 2	1
Elaeocarpaceae	*Elaeocarpus sphaericus* K. Schum.	1
Euphorbiaceae	*Drypetes longifolia* (Bl.) Pax & Hoffm.	1
Euphorbiaceae	*Neoscortechinia forbesii* (Hook. f) Pax ex S. Moore	1
Euphorbiaceae	*Pimeleodendron amboinicum* Hassk.	1
Fabaceae	*Cynometra novoguinensis* Merr. & Perry	1
Fagaceae	*Lithocarpus vinkii* Soepadmo	1
Flacourtiaceae	*Casearia aruensis*	1
Gnetaceae	*Gnetum cuspidatum* Blume	2
Gnetaceae	*Gnetum gnemon*	7
Hymenophyllaceae	*Hymenophyllum*	4
Hymenophyllaceae	*Trichomanes javanica* (Bl.) Sleumer	29
Hymenophyllaceae	*Trichomanes* sp.	27
Icacinaceae	*Gonocaryum litorale*	1

continued

Family	Species	Number
Icacinaceae	*Medusanthera papuana* (Becc.) Howard.	1
Lauraceae	*Beilschmiedia gemiflora* (Bl.) Kosterm.	1
Lauraceae	*Cinnamomum culitlawan* (L.) Kosterm	1
Lauraceae	*Cryptocarya kamahar* Tesch.	2
Lauraceae	*Cryptocarya* sp. 1	4
Lauraceae	*Cryptocarya verrucosa* Teschn.	4
Lauraceae	*Endiandra glauca* R.Br.	7
Lauraceae	*Endiandra* sp. 1	3
Lauraceae	*Litsea firma* Hook. f.	5
Leeaceae	*Leea* sp.	1
Liliaceae	*Dianella* sp.	3
Melastomataceae	*Astronia papetaria* Blume	1
Melastomataceae	*Medinilla octostriata* Ohwi	3
Meliaceae	*Aglaia edulis* (Roxb.) Wall.	3
Meliaceae	*Aglaia* sp. 4	2
Meliaceae	*Chisocheton sageri* (C.DC.) Stevens	1
Meliaceae	*Chisocheton stellatus* Stevens	1
Meliaceae	*Dysoxylum* sp. 1	2
Meliaceae	*Vavaea bantamensis* Koord. & Merr.	4
Moraceae	*Artocarpus communis* J.R. & G. Forst.	1
Moraceae	*Artocarpus vrieseanus* Miq.	1
Moraceae	*Ficus pumila* L.	3
Moraceae	*Ficus punctata* Thunb.	1
Myristicaceae	*Horsfieldia irya* (Gaertn.) Warb.	2
Myristicaceae	*Myristica lauterbachii* Warb.	1
Myrtaceae	*Syzygium longipes* (Warb.) Merr. & Perry	1
Myrtaceae	*Syzygium* sp. 2	1
Myrtaceae	*Syzygium* sp. 3	2
Myrtaceae	*Syzygium* sp. 7	1
Myrtaceae	*Syzygium* sp. 1	1
Myrtaceae	*Syzygium trinerve* (Ridley) Merr. & Perry	1
Nyctaginaceae	*Pisonia longirostris*	10
Pandanaceae	*Frecynetia angustifolia*	6
Pandanaceae	*Freycinetia leptostachya* B.C. Stone	1
Pandanaceae	*Freycinetia linearis* Merr. & Perry	3
Pandanaceae	*Freycinettia leptostachya* B.C. Stone	1
Piperaceae	*Piper abbreviatum* Opiz	4
Piperaceae	*Piper miniatum* Bl.	2

Family	Species	Number
Piperaceae	*Piper* sp. 1	4
Poaceae	*Leptaspis urceolata* (Roxb.) R. Br.	3
Polygalaceae	*Xanthophyllum suberosum* C.T. White	1
Polygalaceae	*Xanthophyllum tenuipetalum* Meyden	11
Polypodiaceae	*Polypodium* sp.	1
Proteaceae	*Helicia hypoglauca* Diels.	1
Proteaceae	*Helicia serrata* (R.Br.) Bl.	2
Rhamnaceae	*Ziziphus horsfieldii* Blume	8
Rubiaceae	*Maschalodesme simplex* Merr.	1
Rubiaceae	*Mastixiodendron pachyclados* Melch.	2
Rubiaceae	*Ophiorrhizza* sp.	3
Rubiaceae	*Urophyllum* sp.	5
Rubiaceae	*Urophyllum umbeliferum* Val.	8
Rutaceae	*Melicope novoguinensis* Val.	1
Sapindaceae	*Guiao* sp.	2
Sapindaceae	*Pometia pinnata* Forst.	47
Sapotaceae	*Madhuca* sp. 1	2
Sapotaceae	*Madhuca* sp. 2	2
Sapotaceae	*Palaquium ridleyi* K. & G.	1
Thelypteridaceae	*Thelypteris* sp.	11
Ulmaceae	*Celtis phillippinensis* Blanco	2
Aclepiadaceae	Unidentified	1
Euphorbiaceae	Unidentified	1
Fern	Unidentified	1
Fern	Unidentified	8
Fern	Unidentified	8
Orchidaceae	Unidentified	1
Orchidaceae	Unidentified	1
Orchidaceae	Unidentified	1
Urticaceae	*Elatostema* sp.	1
Urticaceae	*Elatostema weinlandii* K. Schum	30
Verbenaceae	*Teijsmanniodendron bogoriense* Koorders	2
Unidentified	*Antidesma* sp.	1
Unidentified	*Lyndsaea* sp.	1
Unidentified	*Tectaria* sp.	7

Appendix 9

List of all trees (≥ 1 cm dbh) sampled in 5 Whitaker plots (Furu and Jari) and 2 transects (Tiri) during 2000 RAP surveys, Papua, Indonesia

Yance de Fretes, Ismail A. Rachman, and Elisa Wally

Total area sampled at each site = 0.5 ha. Values indicate number of individuals recorded.

Family	Species	Furu	Tiri	Jari
Actinidiaceae	*Saurauia* sp.		1	
Anacardiaceae	*Campnosperma auriculata*			5
Anacardiaceae	*Campnosperma brevipetiolata*			1
Anacardiaceae	*Mangifera* sp.			1
Anacardiaceae	*Semecarpus forstenii* Bl.	2		
Annonaceae	*Mezzetia* sp.		2	1
Annonaceae	*Mitrephora* sp.		2	
Annonaceae	*Polyalthia discolor* Diels	6		
Annonaceae	*Polyalthia forbesii* F. & M.		1	
Annonaceae	*Polyalthia glauca*			3
Annonaceae	*Polyalthia rumphii* (Bl.) Merr.		1	
Annonaceae	*Polyalthia* sp.	1	1	
Annonaceae	*Popowia* sp.	2	1	
Annonaceae	*Xylopia* sp.			2
Apocynaceae	*Cerbera floribunda*			1
Apocynaceae	*Lepiniopsis ternatensis* Valet.		1	
Aquifoliaceae	*Ilex* sp.		2	
Arecaceae	*Hydriastele* sp.	1		
Arecaceae	*Licuala* sp.			6
Arecaceae	*Pinanga* sp.			2
Arecaceae	*Rhopaloblaste* cf. *brassi* H.E. Moore	10		
Arecaceae	*Rhopaloblaste* sp.			2
Burseraceae	*Canarium acutifolium* (DC.) Merr.		2	
Burseraceae	*Canarium asperum* Benth.	2	3	
Burseraceae	*Canarium denticulatum* Hk.f.	2		
Burseraceae	*Canarium maluense*			7
Burseraceae	*Canarium oleosum*	5		4
Burseraceae	*Canarium* sp.	2		22
Burseraceae	*Haplolobus floribundus*		1	
Burseraceae	*Haplolobus moluccana* H.J.L.	8		

continued

Family	Species	Furu	Tiri	Jari
Burseraceae	*Santiria* sp.			4
Celasteraceae	*Lophopetalum javanicum*		1	
Chrysobalanaceae	*Atuna racemosa* Rafin	4		
Chrysobalanaceae	*Maranthes corymbosa*		1	
Chrysobalanaceae	*Parastemon urophyllus* A.DC.	3		2
Chrysobalanaceae	*Parastemon verstegii* M. & P.	1		
Chrysobalanaceae	*Prunus arborea* (Bl.) Kalkm.	1		
Chrysobalanaceae	*Prunus gazelle-peninsulae* (Kan. & Hatts.) Kalkm.		1	
Chrysobalanaceae	*Prunus schlechteri* (Koehne) Kalkm.	1		
Clusiaceae	*Calophyllum soulatri*			3
Clusiaceae	*Calophyllum* sp.			3
Clusiaceae	*Calophyllum* sp. 1			2
Clusiaceae	*Calophyllum* sp. 2			1
Clusiaceae	*Garcinia celebica* Hall. f.	1	1	7
Clusiaceae	*Garcinia dulcis* (Roxb.) Kurz.	3		
Clusiaceae	*Garcinia forbesii* Lauterb.	4	2	
Clusiaceae	*Garcinia fruticosa* Lauterb.	3		
Clusiaceae	*Garcinia latissima* Miq.		3	
Clusiaceae	*Garcinia* sp.		3	7
Clusiaceae	*Garcinia* sp. 1			3
Clusiaceae	*Garcinia* sp. 2	4		1
Clusiaceae	*Garcinia* sp. 3			1
Clusiaceae	*Garcinia* sp. 4			1
Clusiaceae	*Garcinia rigida* Miq.	1		
Cunoniaceae	*Ceratopetalum succirubrum* C.T. White	20		
Cunoniaceae	*Ceratopetalum succirubrum* C.T. White			
Cunoniaceae	*Schizomeria* sp.	1		
Cunoniaceae	*Teijsmanniodendron aherianum* (Merr.) Bakh.	2		11
Cunoniaceae	*Schizomeria* sp.			2
Cunoniaceae	unidentified			2
Dilleniaceae	*Dillenia quercifolia*			1
Dilleniaceae	*Dillenia* sp.			2
Dipterocarpaceae	*Hopea novoguenensis*	1		7
Dipterocarpaceae	*Vatica rassak* (Korth.) Blume	22	6	
Ebenaceae	*Diospyros discolor*			1
Ebenaceae	*Diospyros herbacarpa* A. Cunn. ex Benth.	1		
Ebenaceae	*Diospyros* sp.		2	1
Ebenaceae	*Diospyros* sp. 1			7
Elaeocarpaceae	*Elaeocarpus* sp. 1	1		1
Elaeocarpaceae	*Elaeocarpus sphaericus* K. Schum	1		
Elaeocarpaceae	*Elaeocarpus stipularis*	1		
Elaeocarpaceae	*Sloanea paradisearum* F. & M.		1	
Euphorbiacea	*Claoxylon polot* (Burm.f.) Merr.		2	
Euphorbiaceae	*Antidesma montanum* Bl.		4	

continued

Family	Species	Furu	Tiri	Jari
Euphorbiaceae	*Aporosa papuana* Pax. & Hoffm.	2		
Euphorbiaceae	*Baccaurea* sp.	1		
Euphorbiaceae	*Blumeodendron elateriospermum* J.J.S.	2		
Euphorbiaceae	*Drypetes longifolia*	7		1
Euphorbiaceae	*Drypetes* sp. 1			3
Euphorbiaceae	*Galearia celebica* Koorders.		1	
Euphorbiaceae	*Glochidion* sp.			2
Euphorbiaceae	*Glochidion* sp. 1			2
Euphorbiaceae	*Macaranga mappa* M.A.		1	
Euphorbiaceae	*Macaranga tesselata* Gage	1		
Euphorbiaceae	*Neoscortechinia*			2
Euphorbiaceae	*Pimeliodendron amboinicum* Hassk.	9	13	4
Euphorbiaceae	unidentified			1
Euphorbiaceae	unidentified		2	
Fabaceae	*Adenanthera microsperma*		1	
Fabaceae	*Adenanthera* sp.			1
Fabaceae	*Archidendron aruensis* (Warb.) de Wit.		3	
Fabaceae	*Cynometra* sp.			7
Fabaceae	*Cynometra novoguinensis* Merr. & Perry	1	2	
Fabaceae	*Cynometra ramiflora* L.		4	
Fabaceae	*Intsia bijuga*		3	
Fabaceae	*Intsia palembanica* Miquel.		4	
Fabaceae	unidentified		1	
Fagaceae	*Lithocarous vinkii* Soepadmo	7		
Fagaceae	*Lithocarpus* sp.		1	
Flacourtiaceae	*Casearia aruensis*		3	
Flacourtiaceae	*Casearia* sp.	1		
Flacourtiaceae	*Flacourtia rukam* Z. & M.		1	
Flacourtiaceae	*Homalium foetidum* (Roxb.) Bth.	1	1	
Flacourtiaceae	*Homalium* sp.			1
Flacourtiaceae	*Ryparosa javanica* (Bl.) Kurz.	2	5	
Gnetaceae	*Gnetum gnemon* L.	16	10	12
Icacinaceae	*Gomphandra australiana* F. & M.	4	6	
Icacinaceae	*Gomphandra montana*			3
Icacinaceae	*Gomphandra* sp. 1		1	
Icacinaceae	*Gomphandra* sp. 2			10
Icacinaceae	*Gonocaryum firiforme* Scheff.		1	
Icacinaceae	*Gonocaryum littorale*		3	
Icacinaceae	*Gonocaryum pyriforme* Scheff.		1	
Icacinaceae	*Gonocaryum* sp. 1			1
Icacinaceae	*Gonocaryum* sp. 2			1
Icacinaceae	*Gonocaryum* sp. 3			3
Icacinaceae	*Medusanthera laxiflora* (Miers.) Howard	1	8	

continued

Family	Species	Furu	Tiri	Jari
Icacinaceae	*Medusanthera papuana* (Becc.) Howard	1	5	
Icacinaceae	*Platea excelsa* Blume	3		3
Icacinaceae	*Pseudobotrys cauliflora* (Pulle) Sleumer		11	
Icacinaceae	*Rhyticaryum oleraqum* Becc.		2	
Icacinaceae	*Stemonurus ammui* (Kaneh.) Sleumer	1		
Icacinaceae	*Stemonurus monticolus* Sleumer	2		
Icacinaceae	*Stemonurus* sp.			2
Lauraceae	*Actinodaphne angustifolia* (Bl.) Nees		1	
Lauraceae	*Actinodaphne procera* (Bl.) Nees	1		
Lauraceae	*Beilschmiedia archboldiana*	1		
Lauraceae	*Beilsmiedia gemmiflora* (Bl.) Kosterm.		2	
Lauraceae	*Cinnamomum cryptocarya laevigata* F. Vill.		1	
Lauraceae	*Cinnamomum culilavan* Bl.		1	
Lauraceae	*Cinnamomum* sp.			1
Lauraceae	*Cryptocarya idenburgensis* Merr.	3		
Lauraceae	*Cryptocarya kamahar* Tesch.	1		
Lauraceae	*Cryptocarya massoy* (Oken.) Kosterm.	1		
Lauraceae	*Cryptocarya palmerensis* Allen		1	
Lauraceae	*Cryptocarya verrucosa* Teschn.	2		
Lauraceae	*Cryptocarya zollingeriana* Miq.	1		
Lauraceae	*Cryptocarya* sp.			6
Lauraceae	*Cryptocarya* sp. 2	1		
Lauraceae	*Cryptocarya* sp. 3		2	
Lauraceae	*Cryptocarya* sp. 4	1		
Lauraceae	*Cryptocarya* sp. 5	1		
Lauraceae	*Cryptocarya* sp. 8	1		
Lauraceae	*Endiandra acuminata* White & Francis	1		
Lauraceae	*Endiandra alleniana* C.T. White	1	1	
Lauraceae	*Endiandra euadenia* (Bl.) Kosterm.	6		
Lauraceae	*Endiandra glauca* R. Br.	1		
Lauraceae	*Endiandra grandifolia* Teschn.	1		
Lauraceae	*Endiandra*	1		
Lauraceae	*Endiandra* sp.		1	1
Lauraceae	*Litsea firma* Bl.	2		
Lauraceae	*Litsea timoriana* Span		4	
Lauraceae	*Litsea* sp.		2	4
Lauraceae	*Phoebe cuneata* Bl.	2		
Lecythidaceae	*Barringtonia racemosa*			3
Lecythidaceae	*Barringtonia racemosa*			1
Lecythidaceae	*Barringtonia* sp.			2
Lecythidaceae	*Barringtonia* sp. 1			4
Lecythidaceae	*Barringtonia* sp. 2	1		
Lecythidaceae	*Barringtonia* sp.	1		

continued

Family	Species	Furu	Tiri	Jari
Lecythidaceae	*Planchonia papuana* Kunth.		1	
Loganiaceae	*Fagraea racemosa*	1	1	2
Melastomataceae	*Astronia papetaria* Bl.	3		
Melastomataceae	*Astronia* sp.			2
Melastomataceae	*Medinila* sp.			2
Melastomataceae	*Memecylon edule* Roxb.	1		
Meliaceae	*Aglaia argentea* Blume		1	
Meliaceae	*Aglaia cucullata* (Roxb.) Pellegr.	2		
Meliaceae	*Aglaia edulis* (Roxb.) Wall.		1	
Meliaceae	*Aglaia sapindina* (F. & M.) Harms.	2		
Meliaceae	*Aglaia* sp.			3
Meliaceae	*Aglaia* sp. 1		2	
Meliaceae	*Aglaia* sp. 2		1	
Meliaceae	*Aglaia* sp. 3		2	
Meliaceae	*Aglaia* sp. 4	1	1	
Meliaceae	*Aglaia sylvestris* (M. Roém.) Merr.	2	5	
Meliaceae	*Chisocheton ceramicus* (Miers.) C.DC.		4	
Meliaceae	*Chisocheton lasiocarpus* (Miq.) Valeton	4	3	
Meliaceae	*Chisocheton patens* Blume		1	
Meliaceae	*Chisocheton stellatus* Stevens	3	11	
Meliaceae	*Chisocheton trichocladus* Harms.		1	
Meliaceae	*Chisocheton* sp.			2
Meliaceae	*Chisocheton* sp. 1	1	1	
Meliaceae	*Dysoxylum*	1		
Meliaceae	*Dysoxylum alliaqum* Bl.	2		
Meliaceae	*Dysoxylum arnoldianum* K. Schum		1	
Meliaceae	*Dysoxylum hexandrum* Merr.	3		
Meliaceae	*Dysoxylum parasiticum*	1		
Meliaceae	*Dysoxylum pettigrewianum* F.M. Bailey		3	
Meliaceae	*Dysoxylum variable* Harm.		1	
Meliaceae	*Dysoxylum* sp.		1	3
Meliaceae	*Dysoxylum* sp. 1	6		5
Meliaceae	*Dysoxylum* sp. 2	1	1	
Meliaceae	*Dysoxylum* sp. 3	2		
Meliaceae	*Trichocladus* Harms.	1		
Meliaceae	*Vavaea bantamensis* Koord. & Merr.	1		
Monimiaceae	*Steganthera* sp.	2		
Moraceae	*Antiaris toxicaria*		1	
Moraceae	*Antiaropsis decipiens*	3		
Moraceae	*Artocarpus elasticus*	1		
Moraceae	*Artocarpus vrieseanus* Miq.	1	1	
Moraceae	*Ficus miquelli* King	1	1	
Moraceae	*Ficus* sp.	1		

continued

Family	Species	Furu	Tiri	Jari
Moraceae	*Ficus* sp. 1	1	2	
Moraceae	*Paratocarpus venenosis* Becc.	1	1	
Moraceae	*Prainea papuana* Becc.	1	1	6
Myristicaceae	*Gymnacranthera paniculata* (A.DC.) Warb.	15	12	1
Myristicaceae	*Gymnacranthera* sp. 1			27
Myristicaceae	*Horsfieldia helwingii* (Warb.) Warb.	1		
Myristicaceae	*Horsfieldia irya* (Gaertn.) Warb.	5		
Myristicaceae	*Horsfieldia kostermansii* J. Sinclair		1	
Myristicaceae	*Horsfieldia* sp.		1	2
Myristicaceae	*Horsfieldia* sp. 1			4
Myristicaceae	*Horsfieldia* sp. 2			2
Myristicaceae	*Myristca tenuivenia* J.Sinclair	1		
Myristicaceae	*Myristica fatua* Hout.	2	1	
Myristicaceae	*Myristica hollrungii* Warb.	7	1	
Myristicaceae	*Myristica kostermansii* J. Sinclair		1	
Myristicaceae	*Myristica lancifolia* Poirret	2		5
Myristicaceae	*Myristica maxima*	1		
Myristicaceae	*Myristica subalunata* Miq.	5		
Myristicaceae	*Myristica sulcata* Warb.	1		
Myristicaceae	*Myristica* sp.	1		17
Myrsinaceae	*Ardisia* sp.		1	
Myrtaceae	*Decaspermum* sp.			1
Myrtaceae	*Syzygium* cf. *furfuraceum* Merr. & Perry	1		
Myrtaceae	*Syzygium geniocalyx* Merr. & Perry		3	
Myrtaceae	*Syzygium leptoneurum* Diels.		2	
Myrtaceae	*Syzygium longipes* (Warb.) Merr. & Perry	2		
Myrtaceae	*Syzygium*		1	
Myrtaceae	*Syzygium* sp.	1		8
Myrtaceae	*Syzygium* sp. 2	1		
Myrtaceae	*Syzygium* sp. 3	2		
Myrtaceae	*Syzygium* sp. 4	2		
Myrtaceae	*Syzygium* sp. 5		1	
Myrtaceae	*Syzygium* sp. 6	1		
Myrtaceae	*Syzygium* sp. 8	1		
Myrtaceae	*Syzygium* sp. 9			1
Myrtaceae	*Syzygium* sp. 10			1
Myrtaceae	*Syzygium* sp. 11			1
Myrtaceae	*Syzygium* sp. 12			2
Myrtaceae	*Syzygium trivene (Ridley)* Merr. & Perry	1		
Myrtaceae	*Syzygium verstegii*		2	
Myrtaceae	unidentified			1
Nyctaginaceae	*Pisonia longirostris* T. et B.	5	2	
Oleaceae	*Chionanthus brassi* (Kobuski) Kiew		1	

continued

Family	Species	Furu	Tiri	Jari
Oleaceae	*Chionanthus oxycarpus* (Lingels) Kiew	1	3	
Oleaceae	*Chionanthus riparius* (Lingels) Kiew	1	1	
Oleaceae	*Chionanthus rupicolus* (Lingels) Kiew	1		
Oleaceae	*Chionanthus sessiliflorus* (Hend. Sh) Kiew.	2	1	
Oleaceae	*Chionanthus*		2	
Oleaceae	*Chionanthus* sp.			2
Pandanaceae	*Pandanus* sp.			3
Podocarpaceae	*Podocarpus blumei*	1		2
Polygalaceae	*Xanthophyllum papuanum* Meyden		1	
Polygalaceae	*Xanthophyllum tenuipetalum* Meyden	8		
Proteaceae	*Helicia moluccana* (R.Br.) Bl.	1		
Proteaceae	*Helicia* sp.			3
Rhamnaceae	*Zyzyphus angustifolia* (Miq.) Hatts.	1	6	
Rosaceae	*Parinarium nonda*			2
Rosaceae	*Parastemon* sp.			1
Rosaceae	*Parinari* sp.			4
Rosaceae	*Prunus* sp.			5
Rosaceae	unidentified			4
Rosaceae/Chrysobalanaceae	*Prunus arborea*			1
Rubiaceae	*Adina multifolia*			7
Rubiaceae	*Aidia racemosa* (Cav.) Triveng	1		
Rubiaceae	*Aidia zippeliana* (Scheff.) Risdale	1		
Rubiaceae	*Lasianthus sylvestris* Bl.	1		
Rubiaceae	*Maschalodesma simplex* Merr.	1	2	
Rubiaceae	*Mastixiodendron pachyclados*		4	5
Rubiaceae	*Mastixiodendron* sp.			2
Rubiaceae	*Nauclea orientalis* L.	2		
Rubiaceae	*Saprosma* sp. 4	1		
Rubiaceae	*Timonius alius* S. Darwin		3	
Rubiaceae	*Timonius novoguinensis* Warb.		1	
Rubiaceae	*Timonius wallichianus* K. Sch.	1		
Rubiaceae	*Timonius* sp.			7
Rubiaceae	unidentified			5
Rubiaceae	unidentified		1	
Rubiaceae	*Urophyllum* sp.	15		
Rubiaceae	*Urophyllum umbeliferum* Val.	8		
Rutaceae	*Tetractomia obovata* Merr.	4		
Rutaceae	*Tetractomia* sp.			4
Santalaceae	*Schropyrum aurantiacum*		1	
Sapindaceae	*Elattostachys zippeliana* Radlk.		2	
Sapindaceae	*Harpulia ramiflora* Radlk.	1		
Sapindaceae	*Jagera serrata* Radlk.	1		
Sapindaceae	*Mischocarpus longifolius*	1		

continued

Family	Species	Furu	Tiri	Jari
Sapindaceae	*Pometia pinnata* Forst.	15	26	2
Sapindaceae	*Pometia* sp.			5
Sapindaceae	unidentified			3
Sapotaceae	*Chrysophyllum fasciculata*			2
Sapotaceae	*Madhuca* sp.	5		
Sapotaceae	*Madhuca* sp. 1	3		
Sapotaceae	*Madhuca* sp. 2	5		
Sapotaceae	*Manilkara fasciculata*			1
Sapotaceae	*Manilkara* sp.			1
Sapotaceae	*Palaquium caophyllum* Pierre		2	
Sapotaceae	*Palaquium* cf. *dasyphyllum* Pierre	1	5	
Sapotaceae	*Palaquium obovatum* Engl.		1	
Sapotaceae	*Palaquium ridleyi* K. & G.	5	2	
Sapotaceae	*Palaquium* sp.		2	3
Sapotaceae	*Palaquium* sp. 1			2
Sapotaceae	*Palaquium* sp. 2			1
Sapotaceae	*Planchonella firma* (Miq.) Dub.	1	1	
Sapotaceae	*Planchonella papuana* Kunth.		2	
Sapotaceae	*Planchonella* sp.		1	1
Sterculiaceae	*Heritiera littoralis* Dryand	1	2	
Sterculiaceae	*Pterygota horsfieldii* Kosterm.	3	1	
Sterculiaceae	*Sterculia* sp.			2
Sterculiaceae	*Sterculia* sp.			1
Theaceae	*Gordonia papuana* Kobuski	1		
Theaceae	*Gordonia* sp.			5
Theaceae	*Ternstroemia cherii* Mart.	1		
Theaceae	*Ternstroemia merilliana* Kobuski	1		
Theaceae	*Ternstroemia toquian* (Blanco) F. Vill.	1		
Theaceae	*Ternstroemia* sp.			2
Thymelaeaceae	*Phaleria microcarpa* (Scheff.) Boerl.		1	
Tiliaceae	*Microcos ceramensis* Burr.	2	4	
Ulmaceae	*Celtis philippinensis*	1	4	2
Ulmaceae	*Celtis* sp.			1
Ulmaceae	*Gironniera hirta* Planch.	1		
Verbenaceae	*Teijsmaniadendron bogoriense* Koordrs.	16	9	11
Verbenaceae	*Teijsmanniodendron aherhianum*			1
Unidentified	unidentified	1		
Unidentified	unidentified			1
Unidentified	unidentified	1		
Unidentified	unidentified	1		
Unidentified	unidentified	1		
	Total Number of Species	**161**	**128**	**118**

Appendix 10

Botanical specimens collected outside of plots during Papua (Indonesia) RAP surveys (2000) and identified by Ismail A. Rachman

Yance de Fretes, Ismail A. Rachman, and Elisa Wally

continued

NO	Family	Species	CI Collection Number	Growth Form	Site JARI	Site FURU	Site TIRI	Voucher Specimen
	PTERIDOPHYTA							
1	Adianthaceae	*Adianthum capillus veneris* L.	596	Terrestrial fern.	x			*
2	Adianthaceae	*Adianthum hispidum* Sw.	597	Terrestrial fern.	x			*
3	Adianthaceae	*Lindsaea* cf. *cultrata* (Willd.) Sm.	561	Terrestrial fern.	x			*
4	Aspleniaceae	*Antrophyum callifolium* Bl.	474	Epiphytic fern.	x			*
5	Aspleniaceae	*Antrophyum semicostatum* Bl.	475	Epiphytic fern.	x			*
6	Aspleniaceae	*Aspenium nidus* L.	600	Epiphytic fern.	x			*
7	Aspleniaceae	*Asplenium bantamense* v.A.v.R.	471	Epiphytic fern.	x			*
8	Aspleniaceae	*Asplenium scortechinii* Bedd.	594	Epiphytic fern.	x			*
9	Aspleniaceae	*Asplenium thunbergii* Kunze	595	Epiphytic fern.	x			*
10	Aspleniaceae	*Dryopteris* cf. *ctenitis* (Kurz.) Ching.	644	Terrestrial fern.	x			*
11	Athyriaceae	*Merinthosorus drynarioides* (Hook.) Copel	560	Epiphytic fern.	x			*
12	Cyatheaceae	*Cyathea* sp.	598	Terrestrial fern 1 m tall.	x			*
13	Hymenophyllaceae	*Hymenophyllum* sp.	557	Terrestrial fern.	x			*
14	Hymenophyllaceae	*Trichomanes grande* Copel	599	Terrestrial fern.	x			*
15	Hymenophyllaceae	*Trichomanes javanica* Bl.	601	Terrestrial fern 5-10 cm	x			*
16	Hymenophyllaceae	*Trichomanes millefolium* Blume	563, 589	Terrestrial fern.	x			*
17	Hymenophyllaceae	*Trichomanes* sp.	593	Terrestrial fern 10-20 cm tall.	x			*
18	Hymenophyllaceae	*Trichomanes* sp. 2	643	Terrestrial fern.	x			*
19	Hymenophyllaceae	*Trichomanes* sp. 4	603	Terrestrial fern.	x			*
20	Lindsaeaceae	*Tapeinidium pinnatum* (Cav.) C.Chr.	650, 680	Terrestrial fern.	x			*
21	Lycopodiaceae	*Lycopodium* sp.	604	Epiphytic fern.	x			*
22	Lyndsaeaceae	*Lindsaea bouillodii* Christ.	556	Hemiepiphytic fern.	x			*
23	Marattiaceae	*Marattia fraxinea* Sm.	605	Terrestrial fern.	x			*
24	Oleandraceae	*Nephrolepis hirsuta* Forst.	467	Terrestrial fern.	x			*
25	Polypodiaceae	*Cyclophorus* sp.	590	Terrestrial fern.	x			*

NO	Family	Species	CI Collection Number	Growth Form	JARI	FURU	TIRI	Voucher Specimen
26	Polypodiaceae	*Diplazium angustifolium* Holt.	559	Terrestrial fern 30-40 cm	x			*
27	Polypodiaceae	*Diplazium* sp.	592	Terrestrial fern.	x			*
28	Polypodiaceae	*Dryopteris* sp.	591	Epiphytic fern.	x			*
29	Polypodiaceae	*Loxogramma involucrata* (D.Doc.) C.	558	Epiphytic fern.	x			*
30	Polypodiaceae	*Loxogramma* sp.	470	Epiphytic fern.	x			*
31	Polypodiaceae	*Microsorium* cf. *punctatum* (L.) Copel	580	Epiphytic fern.	x			*
32	Polypodiaceae	*Phymatosorus nigrescens* (Bl.) PC.	645	Epiphytic fern.	x			*
33	Polypodiaceae	Polypodiaceae 1	602	Epiphytic fern.	x			*
34	Polypodiaceae	Polypodiaceae 2	562	Epiphytic fern.	x			*
35	Polypodiaceae	*Polypodium* sp. 1	473	Epiphytic fern.	x			*
36	Pteridaceae	*Anthropteris palisottii* (Desv.) Alston	634	Terrestrial fern.	x			*
37	Pteridaceae	Fern 5	663	Epiphytic fern.	x			*
38	Pteridaceae	*Tapeinidium luzonicum* (Hook.) Kramer	638	Terrestrial fern.	x			*
39	Schizaeaceae	*Lygodium* sp.	669	Terrestrial fern.	x			-
40	Schizaeaceae	*Schizaea dichotoma* (L.) Sm.	531	Terrestrial fern.	x			*
41	Selaginellaceae	*Selaginella* sp. 1	472	Terrestrial fern.	x			*
42	Selaginellaceae	*Selaginella* sp. 2	607	Terrestrial fern.	x			*
43	Selaginellaceae	*Selaginella* sp. 3	637	Terrestrial fern.	x			*
44	Selaginellaceae	*Selaginella wildenowii* (Desv.) Backer	679	Terrestrial fern.	x			*
45	Vittariaceae	*Vittaria flexuosa* Fec.	609	Epiphytic fern.	x			*
ANGIOSPERMAE			x					
	MONOCOTYLEDONAE							
46	Araceae	*Alocasia* sp.	649	Herb 30-40 cm		x		-
47	Araceae	*Halochlamis beccarii* (Engl.) Engl.	608	Liana up to 15 m tall.	x			-
48	Araceae	*Halochlamis guineensis* Engl. & Krause	532	Liana 5 m tall.	x			-
49	Araceae	*Homalomena* sp.	682	Herb 15-20 cm tall.	x			-
51	Araceae	*Pothos scandens* Linn.	624	Liana 10 m tall.	x			-
52	Araceae	*Rhaphidophora korthalsii* Schott.	682	Liana/creeping herb 8 m tall.	x			-
53	Araceae	*Rhaphidophora novoguinensis* Engl.	612	Liana/creeping herb 4 m tall.	x			-
54	Araceae	*Rhaphidophora* sp. 1	642	Liana/creeping herb 4 m tall.	x			-

continued

NO	Family	Species	CI Collection Number	Growth Form	Site JARI	Site FURU	Site TIRI	Voucher Specimen
55	Araceae	Rhaphidophora sp. 2	673	Liana 12 m tall.	x			-
56	Araceae	Rhaphidophora sp. 3	625	Liana 8 m	x			-
57	Araceae	Rhaphidophora sp. 4	654	Liana/creeping herb 6 m tall.	x			-
58	Araceae	Rhaphidophora sp. 5	630	Liana/creeping herb 15 m tall.	x			-
59	Araceae	Rhaphidophora verstegii Engl. & Krause	533	Herb 10-15 cm tall.	x			-
60	Araceae	Schimatoglottis sp.	684	Small palm 1-2 m tall.	x			-
61	Arecaceae	Arenga microcarpa Becc.	581	Small palm 4 m tall.	x			-
62	Arecaceae	Arenga sp.	683	Small ratan 4 m tall.	x			-
63	Arecaceae	Calamus sp.	631	Small palm 4 m tall.	x			-
64	Arecaceae	Caryota mitis Lour.	685	Palm tree.	x			-
65	Arecaceae	Gronophyllum mayrii (Burr.) H.E. Moore	610	Palm tree.	x			-
66	Arecaceae	Gulubia costata	635	Palm tree.	x			-
67	Arecaceae	Heterospatha sp.	639, 629	Palm tree.	x			-
68	Arecaceae	Hydriastele sp.	534	Palm tree.	x			-
69	Arecaceae	Licuala brevicalyx Becc.	662	Palm tree 6-8 m tall.	x			*
70	Arecaceae	Licuala sp.	611	Palm tree 2 m tall.		x		-
71	Arecaceae	Metroxylon sagu Rottb.	686	Palm tree.	x			-
72	Arecaceae	Orania sp.	688	Palm tree.	x			-
73	Arecaceae	Pigapheta sp.	687	Palm tree.	x			-
74	Arecaceae	Ptychosperma sp.	689	Palm tree.	x			*
75	Arecaceae	Rhopaloblaste cf. brassii H.E. Moore	538	Small palm 4-6 m tall.	x			-
76	Commelinaceae	Forrestia molissima (Bl.) Kds.	568	Herb grass 20-50 cm tall.			x	*
77	Cyperaceae	Mapania sp.	584	Herb 30-40 m tall.	x			-
78	Cyperaceae	Paramapania parvibractea (Clarke) Uitt.	499	Grass 3-5 m tall.	x			*
79	Liliaceae	Dianella sp.	501	Herb 30-50 cm tall.	x			*
80	Liliaceae	Dracaena sp.	613	Shrub 3-4 m tall.	x			-
81	Orchidaceae	Unidentified	615	Epiphytic orchid.	x			*
82	Pandanaceae	Freycinetia excelsa F.v. Mueller	478	Creeping pandanus 6-9 m tall.	x			-
83	Pandanaceae	Freycinetia leptostachya B.C. Stone	575	Creeping pandanus 12 m tall.	x			-
84	Pandanaceae	Freycinetia linearis Merr. & Perry	574	Creeping pandanus 6 m tall.	x	x		-

continued

NO	Family	Species	CI Collection Number	Growth Form	Site JARI	Site FURU	Site TIRI	Voucher Specimen
85	Pandanaceae	*Pandanus* sp. 1	614	Pandanus tree 6 m	x			-
86	Pandanaceae	*Pandanus* sp. 2	583	Small pandanus 1 m tall.	x			*
87	Pandanaceae	*Pandanus tectorius* Soland. Ex Park.	576	Pandanus tree 4-6 m	x			-
DICOTYLEDONAE								
88	Acanthaceae	*Pseuderathemum* sp.	488	Herb 30-40 cm tall.			x	*
89	Actinidiaceae	*Saurauia verstegii* Gilg. ex Lauterb.	489	Shrub 2 m tall.		x		*
90	Actinidiaceae	*Saurauia* sp. 1	585	Shrub 1 m tall.		x		*
91	Actinidiaceae	*Saurauia* sp. 2	572	Shrub 2 m tall.		x		*
92	Anacardiaceae	*Semecarpus aruensis* Engl.	507	Shrub 4 m tall.		x		*
93	Anacardiaceae	*Semecarpus forstenii* Blume	480	Small tree 8 m tall. 14 cm dbh.			x	*
94	Anacardiaceae	*Semecarpus magnificus* K. Schum.	578	Shrub 2 m tall.		x		*
95	Anacardiaceae	*Spondias novoguinensis* Kosterm.	493	Big tree 40 m tall. 60 cm dbh.		x		-
96	Annonaceae	*Goniothalamus* sp.	503	Shrub 3 m tall.		x		*
97	Annonaceae	*Polyanthia celebica* Miq.	582	Small tree.		x		-
98	Annonaceae	*Pseuduvaria* sp. 1	509, 573	Shrub 1 m tall.		x		*
99	Annonaceae	*Pseuduvaria* sp. 2	512	Small tree 5-6 m tall.		x		*
100	Apocynaceae	*Lepiniopsis ternatensis* Valet.	497	Small tree 8 m tall.		x	x	*
101	Apocynaceae	*Ochrosia glomerata* F. & M.	510	Small tree 9 m tall. 12 cm dbh.			x	*
102	Apocynaceae	*Parsonia albo flavescens* (Dennst.) Mabb.	504	Liana 5 m tall.	x			*
103	Asclepiadaceae	*Hoya* sp.	571	Liana 8 m tall.			x	*
104	Asclepiadaceae	*Tylophora cissoides* Blume	579	Liana 8-10 m tall.			x	*
105	Begoniaceae	*Begonia aculeata* Walp.	537, 502	Herb 20-30 cm tall.		x		*
106	Begoniaceae	*Begonia bipinnatifida* J.J.S.	577	Herb 20 cm tall.		x		*
107	Begoniaceae	*Begonia* sp.	536	Herb 15-20 cm tall.		x		*
108	Bignoniaceae	*Tecomanthe dendrophylla* (Bl.) K. Sch.	498	Liana 20 m tall.			x	*
109	Burseraceae	*Canarium acutifolium* (DC.) Merr.	511	Small tree 9 m tall. 10 cm dbh.		x		*
110	Clusiaceae	*Calophyllum inophyllum* L.	541	Tree 15 m tall. 40 cm dbh.	x			*
111	Clusiaceae	*Garcinia forbesii* Pierre	552, 588	Tree 12 m tall. 4 cm dbh.		x		*

continued

NO	Family	Species	CI Collection Number	Growth Form	JARI	Site FURU	TIRI	Voucher Specimen
112	Cucurbitaceae	*Trichosanthes edulis* Rugayah	543	Liana 8 m tall.		x		*
113	Dilleniaceae	*Dillenia quercifolia* Hoogl.	555	Tree 25 m tall. 60 cm dbh.	x			*
114	Ebenaceae	*Diospyros novoguinensis* Bakh.	646	Shrub/small tree 4 m tall.	x			*
115	Elaeocarpaceae	*Elaeocarpus dolichostylis* Schlecht.	542	Tree 20 m tall. 24 cm dbh.	x			-
116	Elaeocarpaceae	*Elaeocarpus glaber* Bl.	587	Tree 15 m tall. 18 cm dbh.	x			-
117	Elaeocarpaceae	*Elaeocarpus petiolatus* Wall.	567	Tree 17 m tall. 16 cm dbh.	x			-
119	Euphorbiaceae	*Claoxylon polot* (Burm.f.) Merr.	554	Shrub 3 m tall.		x		*
120	Euphorbiaceae	*Codiaeum variegatum* (Linn.) Blume	566	Shrub 2-3 m tall.		x		*
121	Euphorbiaceae	*Endospermum moluccanum* (T. & B.) Becc.	545	Tree 12 m tall. 14 cm dbh.		x		-
122	Euphorbiaceae	*Glochidion philipicum* (Cav.) C.B. Robinson	564	Tree 8 m tall. 12 cm dbh.	x			-
123	Euphorbiaceae	*Macaranga fallacina* Pax. & K. Hoffm.	469	Small tree 8 m tall. 5 cm dbh.			x	*
124	Euphorbiaceae	*Macaranga feselata* Gage	551	Small tree 5 m tall. 5 cm dbh.			x	*
125	Euphorbiaceae	*Mallotus floribundus* (Bl.) M.A.	570	Small tree 10 m tall. 12 cm dbh.		x		*
126	Euphorbiaceae	*Pimeleodendron amboinicum* Hassk.	547	Tree 15 m tall. 20 cm dbh.		x		-
127	Fabaceae	*Archidendron aruense* (Warb.) de Wit.	569	Small tree 6 m tall. 6 cm dbh.		x		*
128	Fabaceae	*Desmodium* sp.	586	Herb 30-40 cm	x			-
129	Fabaceae	*Parkia versteeghii* Merr. & Perry	548	Big tree 40 m tall. 60 cm dbh.		x		*
130	Flacourtiaceae	*Casearia aruensis*	491, 565	Shrub 3 m tall.		x		*
131	Flacourtiaceae	*Casearia* sp.	482	Shrub 5 m tall.			x	*
132	Flacourtiaceae	*Scolopia chinensis* (Lour.) Closs	539	Tree 10 m tall. 15 cm dbh.	x			-
133	Gesneriaceae	*Aeschynanthus radicans* Jack	549	Creeping epiphytic herb 2 m tall.		x		*
134	Gesneriaceae	*Cyrtandra sororia* Schltr.	468	Herb 1 m tall.			x	*
135	Gesneriaceae	*Cyrtandra wentiana* Lauterb.	500	Herb 60 cm tall.		x		*
136	Icacinaceae	*Gonocaryum littorale* (Bl.) Sleumer	490	Small tree 8 m tall. 15 cm dbh.		x		*
137	Icacinaceae	*Medusanthera laxiflora* (Miers.) Howard	550, 505	Small tree 8 m tall. 14 cm dbh.		x		*
138	Icacinaceae	*Medusanthera papuana* (Becc.) Howard	513	Small tree 4 m tall.		x		*
139	Icacinaceae	*Pseudobotrys cauliflora* (Pulle) Sleumer	506	Tree 12 m tall.		x		*

continued

NO	Family	Species	CI Collection Number	Growth Form	JARI	Site FURU	TIRI	Voucher Specimen
140	Icacinaceae	*Rhyticaryum oleraceum* Becc.	492, 514	Shrub 3 m tall. 4 cm dbh.		x		*
141	Lauraceae	*Actinodaphne angustifolia* (Bl.) Nees	540	Tree 18 m tall. 20 cm dbh.		x		-
142	Lauraceae	*Cryptocarya brassii* Allen	516	Big tree 25 m tall. 40 cm dbh.		x		*
143	Lecythidaceae	*Barringtonia calyptrocalyx* K. Sch.	546	Small tree 8 m tall. 10 cm dbh.		x		*
144	Leeaceae	*Leea* sp.	515	Shrub 1 m tall.		x		*
145	Loganiaceae	*Fagraea blumei* G. Don	477	Hemiepiphytic liana 4 m tall.			x	-
146	Loganiaceae	*Neuburgia corynocarpa* (C.A. Grey) Leenh.	476	Small tree 9 m tall. 10 cm dbh.		x		*
147	Melastomataceae	*Medinilla angustibasis* Ohwi	692	Shrub.		x		*
148	Melastomataceae	*Medinilla quintuplinervis* Cogn.	620	Shrub 1 m	x			*
149	Meliaceae	*Aglaia argentea* Blume	659, 486	Tree 12 m tall. 14 cm dbh.		x		*
150	Meliaceae	*Aglaia edulis* (Roxb.) Wall.	475	Small tree 12 m tall.		x		*
151	Meliaceae	*Aphanamixis polystachya* (Wall.) Parker	494	Small tree 10 m tall.		x		*
152	Meliaceae	*Chisocheton ceramicus* Miq.	696	Tree 15 m tall.		x		*
153	Meliaceae	*Chisocheton lasiocarpus* (Miq.) Valeton	657	Small tree 9 m tall		x		*
154	Meliaceae	*Dysoxylum alatum* Harms.	495	Small tree 7 m tall.		x		*
155	Meliaceae	*Dysoxylum pettygrewianum* F.M. Bailey	579	Tree 10 m tall.		x		-
156	Meliaceae	*Dysoxylum variabile* Harms.	670	Tree 10 m tall.		x		*
157	Menispermaceae	*Macrococulus pomiferus* Becc.	496	Liana 6 m tall.	x			-
158	Moraceae	*Antiaropsis decipiens* K. Schum.	517	Shrub 4-5 m tall. 4cm dbh.			x	*
159	Moraceae	*Ficus damaropsis* Diels.	616	Small tree 6 m tall.	x			-
160	Moraceae	*Ficus* sp.	690	Tree 8 m tall. 12 cm dbh.	x			*
161	Moraceae	*Ficus* sp. 1	647		x			*
162	Moraceae	*Ficus* sp. 2	695		x			*
163	Moraceae	*Ficus* sp. 3	655			x		*
164	Moraceae	*Ficus subulata* Bl.	691	Shrubling 12 m tall.	x			-
165	Moraceae	*Prainea papuana* Becc.	617	Small tree 10 m tall. 12 cm dbh.	x			*
166	Myristicaceae	*Gymnacranthera paniculata* (A.DC.) Warb.		Tree 18 m tall. 20 cm dbh.	x			-

continued

NO	Family	Species	CI Collection Number	Growth Form	JARI	Site FURU	TIRI	Voucher Specimen
167	Myristicaceae	Horsfieldia irya Warb.	518	Small tree 10 m tall. 12 cm dbh.			x	*
168	Myristicaceae	Myristica lancifolia Poirret	664	Small tree 8 m tall. 10 cm dbh.		x		*
169	Myristicaceae	Myristica lauterbachiana Warb.	519, 658	Small tree 7 m tall. 6 cm dbh.		x		*
170	Myristicaceae	Myristica subalunata Miq.	522, 529	Small tree 9 m tall. 10 cm dbh.		x		*
171	Myrsinaceae	Ardisia sp.	619	Shrub 4 m tall.	x			-
172	Myrsinaceae	Fittingia sp. 1	523	Shrub 1 m tall.		x		*
173	Myrsinaceae	Fittingia sp. 2	518	Shrub 1 m tall.		x		*
174	Myrsinaceae	Maesa sp.	524	Shrub 8 m tall.	x			*
175	Myrtaceae	Decaspermum neurophyllum Lauterb. ex Sch.	530	Shrub 4 m tall. 5 cm dbh.		x		*
176	Myrtaceae	Syzygium acorantha Diels.	520	Small tree 8 m tall. 10 cm dbh.		x		*
177	Myrtaceae	Syzygium longipes (Warb.) Merr. & Perry	553	Tree 12 m tall. 14 cm dbh.		x		*
178	Myrtaceae	Syzygium versteghii Lauterb.	656			x		-
179	Myrtaceae	Tristania whiteana Griff.	648	Tree 10 m tall.	x			*
180	Nyctaginaceae	Pisonia longirostris T. et B.	525	Shrub 3-4 m tall.		x		*
181	Pentaphragmaceae	Pentaphragma grandiflora	521	Herb 80-100 cm tall.			x	*
182	Piperaceae	Piper decumanum (Rumph.) L.	640	Liana 6 m tall.			x	-
183	Piperaceae	Piper mestonii F.M. Bailey	694, 641	Liana 10 m tall.		x		*
184	Pittosporaceae	Pittosporum sinuatum Bl.	653	Shrub 1.5-2 m tall.		x		*
185	Podocarpaceae	Nageia wallichiana (Presl.) Del.	675	Small tree 8 m tall. 6 cm dbh.	x			-
186	Podocarpaceae	Podocarpus neriifolius D.Don.	622	Big tree 30 m tall. 40 cm dbh.	x			-
187	Polygalaceae	Xanthophyllum suberosum C.T. White	652				x	-
188	Rhizophoraceae	Anisophyllea sp.	678	Tree	x			-
189	Rhizophoraceae	Gynotroches axillaris Bl.	629	Tree 15 m tall. 14 cm dbh.	x			*
190	Rosaceae	Prunus sp.	661	Small tree 14 m tall. 12 cm dbh.		x		*
191	Rubiaceae	Gardenia sp.	672	Shrub 8 m tall. 10 cm dbh.		x		-
192	Rubiaceae	Unidentifed	665	Shrub 8 m tall. 10 cm dbh.		x		*

continued

NO	Family	Species	CI Collection Number	Growth Form	JARI	FURU	TIRI	Voucher Specimen
						Site		
193	Rubiaceae	Ixora kochii Brem	628	Shrub 4-5 m tall. 6 cm dbh.	x			*
194	Rubiaceae	Maschalodesme simplex Merr.	667	Shrub 1-2 m tall.		x		*
195	Rubiaceae	Mussaenda cylindrocarpa Burck.	668, 623	Shrub 6 m tall.	x		x	*
196	Rubiaceae	Mycetia javanica (Bl.) Reinw.	636	Herb 70 cm tall.		x		-
197	Rubiaceae	Myrmecodia lanceolata Val.	675	Epiphytic herb.	x			*
198	Rubiaceae	Psychotria sp.	633	Shrub 2 m	x			-
199	Rubiaceae	Timonius novoguinensis Warb.	677	Small tree 10 m tall.		x		*
200	Rubiaceae	Uncaria longiflora (Poir.) Merr.	635	Liana 20 m tall.	x			-
201	Rubiaceae	Urophyllum umbelliferum Val.	671	Small tree 5 m tall. 5 cm dbh.		x		*
202	Rutaceae	Melicope novo-guinensis Val.	660	Shrub 6 m tall. 6 cm dbh.	x			*
203	Rutaceae	Monanthocitrus aruensis	666	Shrub 5 m tall. 3 cm dbh.		x		*
204	Santalaceae	Scleropyrum aurantiacum (Lauterb. & K. Schum) Polger	621	Small tree 5 m tall. 5 cm dbh.	x	x		*
205	Sapindaceae	Euphorianthus sp.	627		x			-
206	Sapindaceae	Harpulia ramiflora Radlk.	693	Small tree 10 m tall. 12 cm dbh.		x		*
207	Sapindaceae	Jagera serrata Radlk.	651	Tree 16 m tall. 15 cm dbh.	x			*
208	Sapindaceae	Rhysotoechia sp.	483	Shrub 4 m tall.	x			-
209	Sapotaceae	Palaquium sp.	508	Small tree 5 m tall.	x			*
210	Theaceae	Ternstroemia cherii Mart.	485	Tree 25 m tall.	x			*
211	Thymelaeaceae	Phaleria macrocarpa (Scheff.) Boerl.	674	Small tree 13 m tall. 12 cm dbh.		x		*
212	Urticaceae	Elatostema parasiticum Bl.	526	Herb 1 m tall.		x		*
213	Urticaceae	Elatostema sp. 1	632	Herb 40-60 cm tall.		x		-
214	Urticaceae	Elatostema cf. weinlandii K. Schum	466	Herb 40-60 cm tall.		x		*
215	Verbenaceae	Clerodendrum buruana Miq.	527	Shrub 3 m tall.		x		*
216	Winteriaceae	Zygogynum calothyrsum (Diels) Vink	676, 528	Shrub 3-5 m tall.		x	x	*

* Fertile specimen deposited in Herbarium Bogoriense.

- Sterile specimen

Appendix 11

Plant species grown in gardens and used by the Dabra community (Mamberamo Basin), Papua, Indonesia

Yance de Fretes, Ismail A. Rachman, and Elisa Wally

Family	Species	Growth Form, Uses
Acanthaceae	*Abelmoschus manihot* (L.) Medik.	Shrub, Vegetable
Acanthaceae	*Gendarussa vulgaris* Nees	Ornamental, Medicinal
Anacardiaceae	*Mangifera* spp.	Tree, Fruit
Anacardiaceae	*Spondias cytherea* Sonn.	Tree, Fruit
Annonaceae	*Annona muricata* L.	Fruit
Apocynaceae	*Allamanda catharica* L.	Woody liana, Ornamental
Apocynaceae	*Vinca rosea* L.	Herb, Ornamental
Araceae	*Caladium bicolor* (W. Ait.) Vent.	Ornamental
Araceae	*Calocasia esculenta* (L.) Schott.	Food
Araceae	*Colocasia esculenta* (L.) Schott.	Herb, Food
Araceae	*Xanthosoma sagittifolium* Schott.	Herb, Food
Arecaceae	*Areca cathecu* L.	Tree, Fruit
Arecaceae	*Cocos nucifera* L.	Tree, Fruit, Vegetable
Arecaceae	*Salacca zalacca* (J. Gaertner) Voss ex Vilmorin	Fruit
Asteraceae	*Cosmos caudatus* H.B.K.	Ornamental
Asteraceae	*Tagetes erecta* L.	Herb, Ornamental
Brommeliaceae	*Ananas comoscus* (L.) Merr.	Herb, Fruit
Caricaceae	*Carica pepaya* L.	Herb, Fruit
Convolvulaceae	*Ipomoea aquatica* Forst.	Herb, Vegetable
Convolvulaceae	*Ipomoea batatas* (L.) L.	Herb, Food
Euphorbiaceae	*Acalypha siamensis* Oliv.	Hedge, Ornamental
Euphorbiaceae	*Codiaeum variegatum* (L.) Bl.	Shrub, Ornamental
Euphorbiaceae	*Jatropha curcas* L.	Small tree, Ornamental, Medicinal
Euphorbiaceae	*Manihot esculenta* Crantz.	Shrub, Food
Euphorbiaceae	*Pedilanthus tithymaloides* Poit.	Ornamental
Fabaceae	*Arachis hypogaea* L.	Herb, Food
Fabaceae	*Vigna unguiculata* (L.) Walp.	Twining Herb, Vegetable
Flacourtiaceae	*Pangium edule* Reinw.	Tree, Vegetable
Lamiaceae	*Orthosiphon stamineus* Benth.	Herb, Ornamental, Medicinal
Lamiaceae	*Plectranthus scutellaroides* (L.) R.Br.	Herb, Ornamental
Liliaceae	*Belamcanda chinensis* (L.) DC.	Herb, Ornamental

continued

Family	Species	Growth Form, Uses
Maranthaceae	*Phacelophyrynium maximum* (Bl.) K. Schum.	Roof construction
Moraceae	*Artocarpus communis* Forst.	Vegetable, Food
Moraceae	*Artocarpus integer* (Thunb.) Merr.	Tree, Fruit
Musaceae	*Musa* spp.	Herb, Fruit
Myrtaceae	*Psydium guajava* L.	Small tree, Fruit
Myrtaceae	*Syzygium aqueum* (Burm.f.) Alston	Tree, Fruit
Nyctaginaceae	*Bougainvillea spectabilis* Willd.	Woody liana, Ornamental
Oxalidaceae	*Averhoa carambola* L.	Tree, Fruit
Piperaceae	*Piper betle* L.	Liana, Chewing
Poaceae	*Zea mays* L.	Food
Rubiaceae	*Coffea robusta* Linden	Shrub, Beverage
Rubiaceae	*Gardenia jasminoides* Ellis	Shrub, Ornamental
Rubiaceae	*Morinda citrifolia* L.	Tree, Fruit, Medical
Rutaceae	*Citrus nobilis* Lour.	Tree, Fruit
Sapindaceae	*Pometia pinnata* J.R. & G. Forst.	Tree, Fruit
Sterculiaceae	*Theobroma cacao* L.	Tree, Fruit, Beverage
Verbenaceae	*Clerodendrum paniculatum* L.	Shrub, Ornamental
Verbenaceae	*Duranta erecta* L.	Shrub, Ornamental
Zingiberaceae	*Acorus calamus* L.	Herb, Medicinal
Zingiberaceae	*Alpinia galanga* (L.) Swartz.	Herb, Spice, Medicinal
Zingiberaceae	*Curcuma domestica* Sw.	Herb, Food, Spice
Zingiberaceae	*Zingiber officinalis* Rosc.	Herb, Spice, Medicinal

Appendix 12

Sampling stations for aquatic insect surveys in the Dabra area, Papua, Indonesia

Dan A. Polhemus

Station 1: Upper Furu River (trib. to Idenburg River), 3 km. SE of Dabra, 90 m., water temp. 25.5°C, 4–7 September 2000, 3°17'04"S, 138°38'10"E.

Station 2: Swampy area in forest near Furu River (trib. to Idenburg River), 3 km. SE of Dabra, 80 m., 4 September 2000, 13:00–13:15 hrs, 3°16'45"S, 138°38'24"E.

Station 3: Cascading tributary to upper Furu River (trib. to Idenburg River), 3 km. SE of Dabra, 90–110 m., water temp. 25°C, 5 September 2000, 06:30–11:00 hrs, 3°17'04"S, 138°38'10"E.

Station 4: Furu River (trib. to Idenburg River) and sand bottomed tributary, 3 km. SE of Dabra, 80 m., water temp. 25.5°C, 7 September 2000, 08:30–09:30 hrs, 3°16'45"S, 138°38'24"E.

Station 5: Doorman River just upstream from confluence with Idenburg River, W of Dabra, 50 m., water temp. 26.5°C (main river), 29°C (side pools), 6 September 2000, 10:30–12:30 hrs, 3°14'40"S, 138°35'29"E.

Station 6: Tiri River (trib. to Doorman River), 4.5 km. SW of Dabra, 80 m., water temp. 25°C, 9–13 September 2000, 3°17'30"S, 138°34'53"E.

Station 7: Pools in sandy overflow channel branching from Tiri River (trib. to Doorman River), 4.5 km. SW of Dabra, 80 m., water temp. 25°C, 9–13 September 2000, 3°17'30"S, 138°34'53"E.

Station 8: Forest stream (trib. to Tiri River), 2.75 km. W of Dabra, 50 m., water temp. 25°C, 14 September 2000, 08:00–08:30 hrs, 3°16'18"S, 138°35'24"E.

Station 9: Rocky hill forest streamlet, 2.5 km. W of Dabra, 90 m., water temp. 25.5°C, 14 September 2000, 09:00–10:30 hrs, 3°16'07"S, 138°35'30"E.

Appendix 13

Aquatic insects captured at nine sampling stations in the Dabra area, Papua, Indonesia

Dan A. Polhemus

Each station is described in Appendix 12. Where an exact species determination has not been made, the following notations are employed:

n. sp.: indicates that the species is clearly new to science. In cases where multiple new species in the same genus were present, as in *Microvelia* and *Rhagovelia*, a numbering system was used (ie., "n. sp. #1"). The numbering is consistent for a given taxon throughout the tables in this report.

sp. undet.: indicates that the species has not yet been definitively identified, and may possibly be undescribed, or simply unidentifiable given the limitations of the current taxonomic literature.

Sampling Stations	FURU RIVER				TIRI RIVER					Stations Combined	
	1	2	3	4	5	6	7	8	9	Furu	Tiri
HETEROPTERA											
Corixidae											
Micronecta sp. undet.							X			-	X
Gerridae											
Iobates affinis	X		X	X		X	X	X		X	X
Limnometra kallisto	X	X	X			X	X			X	X
Limnogonus darthulus				X						X	-
Metrobatoides n. sp.					X	X				-	X
Metrobatopsis flavonotatus				X			X	X		X	X
Neogerris parvulus		X								X	-
Ptilomera aello	X					X			X	X	X
Tenagogonus sp. undet. #1	X			X		X	X	X		X	X
Tenagogonus sp. undet. #2							X			-	X
Hebridae											
Hebrus n. sp. #1	X					X				X	X
Hebrus n. sp. #2						X	X			-	X
Hydrometridae											
Hydrometra sp. undet.				X			X			X	X
Mesoveliidae											
Mesovelia subvittata	X					X				X	X
Mesovelia melanesica				X						X	-

continued

Sampling Stations	FURU RIVER				TIRI RIVER					Stations Combined	
	1	2	3	4	5	6	7	8	9	Furu	Tiri
Naucoridae											
Aptinocoris ziwa	X					X				X	X
Idiocarus papuus	X				X	X				X	X
Idiocarus minor	X									X	-
Sagocoris biroi						X				-	X
Cavocoris sp. undet.	X					X				X	X
Notonectidae											
Anisops sp. undet.						X				-	X
Enithares sp. undet.							X			-	X
Ochteridae											
Ochterus sp. undet. #1	X									X	-
Ochterus sp. undet. #2						X	X			-	X
Ochterus sp. undet. #3						X				-	X
Veliidae											
Microvelia n. sp. #1		X								X	-
Microvelia n. sp. #2	X			X		X	X			X	X
Microvelia n. sp. #3	X			X		X				X	X
Microvelia n. sp. #4	X		X					X	X	X	X
Microvelia n. sp. #5						X	X			-	X
Neusterensifer n. sp. #1	X		X	X			X			X	X
Neusterensifer n. sp. #2	X		X							X	-
Neusterensifer n. sp. #3						X				-	X
Rhagovelia n. sp. #1			X						X	X	X
Rhagovelia n. sp. #2	X		X	X		X				X	X
Rhagovelia n. sp. #3				X		X	X	X		X	X
Rhagovelia n. sp. #4	X		X							X	-
Rhagovelia n. sp. #5					X					-	X
Rhagovelia n. sp. #6						X				-	X
Strongylovelia sp. undet.	X		X	X		X	X		X	X	X
ZYGOPTERA											
Calopterygidae											
Neurobasis australis	X			X		X				X	X
Chlorocyphidae											
Rhinocypha tincta amanda	X					X				X	X
Coenagrionidae											
Agriocnemus ensifera		X								X	-
Papuagrion occipitale	X									X	-
Pseudagrion farinicolle				X						X	-
Pseudagrion civicum				X		X				X	X
Palaiagria ceyx						X				-	X
Platycnemididae											
Idiocnemis obliterata	X									X	-

continued

Sampling Stations	FURU RIVER				TIRI RIVER					Stations Combined	
	1	2	3	4	5	6	7	8	9	Furu	Tiri
Protoneuridae											
Nososticta nigrofasciata	X			X		X	X	X		X	X
Nososticta melanoxantha							X			X	X
Nososticta beatrix						X				-	X
Nososticta irene	X									X	-
Nososticta erythrura						X				-	X
Selysioneura phasma	X									X	-
COLEOPTERA											
Gyrinidae											
Merodineutes sp. undet.						X				-	X
Macrogyrus sp. undet.			X						X	X	X
Total Heteroptera	19	3	9	12	3	23	15	5	4	27	32
Total Zygoptera	7	1	0	4	0	7	2	1	0	11	8
Total Coleoptera	0	0	1	0	0	1	0	0	1	1	2

Appendix 14

Annotated checklist of aquatic insects collected during the Dabra RAP survey, Papua, Indonesia.

Dan A. Polhemus

HETEROPTERA

Gerridae

Iobates affinis Esaki
Taken on flowing midstream pools.

Limnometra kallisto (Kirkaldy)
Common in all standing water habitats.

Limnogonus darthulus (Kirkaldy)
Present on pools of the sand bottomed tributary to the Furu River.

Metrobatoides n. sp.
Taken from eddies and side pools at the lower Doorman River, and from flowing pools on the Tiri River.

Metrobatopsis flavonotatus Esaki
In quiet backwaters, sometimes intermixed with *I. affinis*.

Neogerris parvulus (Stål)
Taken only in a swampy area near the Furu River (same habitat as *Microvelia* n. sp. #1).

Ptilomera aello Breddin
Common on all flowing streams in the area.

Tenagogonus sp. undet. #1
Taken adjacent to the Furu River on a side pool with a substrate of leaf litter. Also common on the backwaters of the Tiri River, particularly on standing pools in the sandy overflow channels.

Tenagogonus sp. undet. #2
Taken intermixed with *Tenagogonus* sp. undet. #1 on the pools in the sandy Tiri River overflow channel.

Veliidae

Microvelia n. sp. #1
A brown, sexually dimorphic species with a short pronotum, taken from shallow pools in forest swamp near Furu River.

Microvelia n. sp. #2
An active, black species with white spotted wings, found on wet midstream rocks, and quick to take flight when disturbed.

Microvelia n. sp. #3
A black species with a red head and narrow body. Taken from a side pool with a substrate of leaf litter adjacent to the Furu River, from the sandy tributary to the Furu River, and from a small tributary with cobbles far upstream on the Tiri River (at this latter site a single male was taken from flowing pocket water amid cobbles).

Microvelia n. sp. #4
An elongate brown species with pale legs. A good series was taken from a hill streamlet behind Dabra, a few others on a cascading tributary to the Furu River, and one specimen in a side pool near the Furu River.

Microvelia n. sp. #5
A large, sexually dimorphic species with frosty grey females.

Neusterensifer n. sp. #1
This dark brown species was common on pocket water along stream margins and also occurred on pools in the Tiri River overflow channel.

Neusterensifer n. sp. #2
This orange-brown species was taken from shallow, shaded pools on a side channel to the Furu River.

Neusterensifer n. sp. #3
This brown species was taken from damp, muddy areas on the forest floor near the Tiri River.

continued

Rhagovelia n. sp. #1
An orange and black species present on first order hill streams.

Rhagovelia n. sp. #2
A robust black species in the *novacaledonica* group, present on rocky streams.

Rhagovelia n. sp. #3
A black species with a narrow body, present on sandy streams.

Rhagovelia n. sp. #4
A black and white species present on very shallow laminar flows over sandy areas in the deep forest.

Rhagovelia n. sp. #5
A large grey species in the *caesius* group, taken along the margins of the lower Doorman River.

Rhagovelia n. sp. #6
A smaller black species in the *caesius* group taken from the Tiri River.

Strongylovelia sp. undet.
Common on still pools and stream backwaters.

Mesoveliidae
Mesovelia melanesica
One pair of this species was taken from a sandy tributary to the Furu River.

Mesovelia subvittata
Taken on wet mud banks and damp areas on the forest floor.

Hebridae
Hebrus n. sp. #1
A dark species with a reddish head. Taken from midstream rocks on the main channel above Furu River camp, and further specimens were obtained from wet logs at the Tiri River locality.

Hebrus n. sp. #2 (nr. *papuanus* Horvath)
This is a larger dark species with a white fascia on hemelytra. Taken from wet logs in pools along the sandy overflow channel at Tiri River.

Hydrometridae
Hydrometra sp. undet.
A small series was taken at the sandy tributary to Furu River, and another larger series from pools in the sandy overflow channel off the Tiri River.

Corixidae
Micronecta sp. undet.
An attractive little pale tan species with longitudinal dark lines on the hemelytra. Taken from the benthic zone of standing pools in the sandy overflow channel at Tiri River, in water 15–30 cm deep.

Naucoridae
Aptinocoris papuus Montandon
Particularly abundant amid small rocks and gravel lying over sand in areas where water dropped from small sandbar deposits into slightly deeper runs.

Cavocoris sp. undet. (possibly *bisulcus* La Rivers)
Found only in shallow riffles amid medium sized cobbles lying on top of sand in areas of swift flow. Not taken from stream margins in pocket water, slow pools, or amid root mats.

Idiocarus papuus Polhemus & Polhemus
Taken in swift midstream riffles.

Idiocarus minor La Rivers
Not as common as the larger *I. papuus*. Taken in the fast water of midstream riffles.

Sagocoris biroi Montandon
Taken along riffle margins at Tiri River. Not taken in midstream sections, or in water over 30 cm deep.

Ochteridae
Ochterus sp. undet. #1
A small series was taken by pyrethrin fogging of a wet vertical mud bank along the Furu River below camp.

Ochterus sp. undet. #2
A maculate brown species fogged from wet logs bordering pools in the sandy overflow channel off the Tiri River.

Ochterus sp. undet. #3
Taken from seasonally flooded mud pans on the forest floor near the Tiri River (same habitat as *Neusterensifer* sp. 3).

Notonectidae
Anisops sp. undet.
One specimen was taken in the pools along the sandy overflow channel at Tiri River. All others (a large series) were taken from an isolated pool in a kettle-like depression on the raised forest bank of the Tiri River upstream from camp. This is the same spot that produced the only patrolling male of *Nososticta erythrura* and demonstrates how *Anisops* require a permanent, uniformly fish-free, standing water environment. As a result they are localized at any given site, although their good powers of dispersal allow them to colonize widely scattered suitable habitats.

continued

Enithares sp. undet.

A small species, present in low numbers in the uppermost pools along the sandy side channel at Tiri River. Both immatures and adults were present, indicating that the species is breeding in these habitats. As with *Anisops*, the absence of fish (and possibly current) seems to be necessary for the presence of this species.

ZYGOPTERA

Calopterygidae

Neurobasis australis Selys

This is a large, metallic green damselfly that is widely distributed in southern New Guinea. Individuals flew conspicuously along stream corridors, often perching on overhanging branches or prominent streamside rocks.

Chlorocyphidae

Rhinocypha tincta amanda Lieftinck

This moderately small, robust-bodied damselfly with dark, metallic-colored wings is one of the most widespread aquatic insects in New Guinea below 1,000 m elevation. It was found at nearly all stream sites surveyed.

Coenagrionidae

Agriocnemus sp. (prob. *ensifera*)

A few individuals were observed above a shallow flooded, grassy ditch at Dabra, but it was not possible to collect any specimens.

Palaiargia ceyx Lieftinck

A single female was taken from far up the Tiri River when it flew down a small, cobble-bottomed tributary and perched on a rock. No male was seen but the specimen in hand matches Lieftinck's (1949) description.

Papuagrion occipitale (Selys)

Females of this large, slender, olive green species were taken flying under the tarps at Furu River camp.

Pseudagrion farinicolle Lieftinck

A single male was taken flying along the sandy tributary at the Furu River; not seen elsewhere.

Pseudagrion civicum Lieftinck

This moderate sized, yellow and black damselfly is very widespread, occurring in slow sections of streams throughout lowland New Guinea. In the Dabra area it was frequently observed flying low over the water next to vertical stream banks with overhanging vegetation.

Platycnemididae

Idiocnemis obliterata Lieftinck

A single individual was taken above a shallow laminar flow of water over sand in the forest next to the Furu River. This species was difficult to follow in the low light of the forest, and their habit of flying very low above a rivulet next to overhanging vegetation made them difficult to capture.

Protoneuridae

Nososticta beatrix (Lieftinck)

A small metallic blue damselfly with a yellow abdomen tip, found hovering about 2–10 m above the ground in patches of sunlight in forest clearings. Several individuals frequently occupied the same light column. Never seen along streams.

Nososticta erythrura (Lieftinck)

Males of this species are generally dark colored with a red abdomen tip. Only a single specimen was taken, flying in dappled sunlight above a water filled depression adjacent to the Tiri River.

Nososticta irene (Lieftinck)

A single female of this species was taken flying next to a tall, shaded bank along the inside of a stream bend on the Furu River below camp. Based on data collected by the 3rd Archbold Expedition (Lieftinck 1949) this appears to be a species of rocky streams in the foothill zone that barely strays into the lower elevational range covered by the present RAP survey.

Nososticta melanoxantha (Lieftinck)

An orange and black species found at pools along the sandy overflow channels, usually where they narrowed out next to a dark bank or downed log. Typically only one male was found per pool, hovering 1.0–1.5 m above the surface.

Nososticta nigrofasciata Lieftinck

This blue and black species is rather widely distributed in New Guinea, occurring in lowland forests along the north and south coasts, and on Biak Island. It was extremely abundant along the sand bottomed overflow channel that branched from the Tiri River below the second camp. A few individuals were also found along the Furu River and Tiri River, hovering around wet logs in the stream channel. Females coupled in tandem pairs were observed ovipositing on wet logs in the sandy side channels at Tiri River. A pale blue and black form that seems morphologically indistinguishable from this species was flying in the forest understory, usually around clearings with a litter of downed branches. Females were observed perching on downed wood in these habitats and may oviposit there, although this form may also utilize the intermittently flooded mud pans on the forest floor that also harbored *Neusterensifer* n. sp. #3. If so,

continued

the immatures may hatch out and feed on *Anopheles* mosquito larvae in the rainy season. Adults fly fairly low, often less than 1 m off the ground.

Selysioneura phasma Lieftinck
Steve Richards took one specimen at the Furu River.

COLEOPTERA

Gyrinidae
Merodineutes sp. undet.
A few individuals were present on the main stream at Furu River, and this species was common along the margins of stream pools at the Tiri River.

Macrogyrus sp. undet.
A series was taken on a steep, rocky streamlet in the hills behind Dabra, on small pools below the trail crossing.

LITERATURE CITED

Lieftinck, M.A. 1949. The dragonflies (Odonata) of New Guinea and neighboring islands. Part VII. Results of the Third Archbold Expedition 1938-1939 and of the Le Roux Expedition 1939 to Netherlands New Guinea (II. Zygoptera). Nova Guinea, New Series I: 1–82.

Appendix 15

Aquatic insects collected in the Cyclops Mountains, Papua, Indonesia

Dan A. Polhemus

COLLECTION SITES:

Station C1 Danyamo Creek, 2.7 km upstream from Yongsu Camp, water temp. 25°C, pH 8.5, 28 August 2000, 2°27'25" S, 140°29'29" E (G. A. Allen collector).

Station C2 Brackish pools behind beach at Yongsu Camp, water temp. 28.5°C, 30 August 2000, 2°26'06" S, 140°29'05" E (G. A. Allen collector).

Station C3 Ornamental hotel pools at Sentani, 20 m (70 ft.), 17 September 2000, 14:00–14:30 hrs, 2°34'22" S, 140°32'07" E.

Station C4 Rocky stream above Pos Tujuh, NW of Sentani, 260–300 m, water temp. 22.5°C, 18 September 2000, 14:00–16:30 hrs, 2°32'26" S, 140°30'47" E.

Station C1	Station C2	Station C3	Station C4
HETEROPTERA	**HETEROPTERA**	**HETEROPTERA**	**HETEROPTERA**
Gerridae	Gerridae	Gerridae	Gerridae
Ptilomera cheesmanae Hungerford & Matsuda	*Limnometra ciliata* Mayr	*Limnogonus* sp. undet.	*Ptilomera cheesmanae* Hungerford & Matsuda
Veliidae		Notonectidae	Hebridae
Neusterensifer sp. undet. (*cyclops?*)		*Anisops* sp. undet.	*Hebrus* sp. undet.
Rhagovelia sp. undet.		Veliidae	Naucoridae
		Microvelia sp. undet.	*Cavocoris* sp. undet.
			Idiocarus sp. undet.
			Tanycricos n. sp.
			Notonectidae
			Enithares sp. undet.
			Ochteridae
			Ochterus n. sp.
			Veliidae
			Microvelia sp. undet. #1
			Microvelia sp. undet. #2
			Neusterensifer sp. undet.
			Rhagovelia sp. undet. #1
			Rhagovelia sp. undet. #2

Appendix 16

List of butterflies collected around Yongsu Dosoyo, Papua, Indonesia

Edy Rosariyanto, Henk van Mastrigt, Henry Silka Innah, and Hugo Yoteni

Family	Subfamily	Species	Forest	Garden/ Forest Edge
Papilionidae	Papilioninae	*Atrophaneura polydorus*	X	X
Papilionidae	Papilioninae	*Ornithoptera priamus*	X	X
Papilionidae	Papilioninae	*Graphium agamemnon*	X	-
Papilionidae	Papilioninae	*Graphium euryplus*	-	X
Papilionidae	Papilioninae	*Papilio aegeus*	X	X
Papilionidae	Papilioninae	*Papilio euchenor*	X	X
Papilionidae	Papilioninae	*Papilio ulysses*	X	X
Pieridae	Pierinae	*Catopsilia pomona*	X	X
Pieridae	Pierinae	*Eurema candida*	X	X
Pieridae	Pierinae	*Eurema hecabe*	-	X
Pieridae	Pierinae	*Eurema blanda*	-	X
Pieridae	Pierinae	*Elodina andropis*	-	X
Pieridae	Pierinae	*Saletara cycinna*	-	X
Pieridae	Pierinae	*Appias ada*	-	X
Lycaenidae	Riodeninae	*Dicalleneura decorata*	-	X
Lycaenidae	Lycaeninae	*Arhopala aexone*	-	X
Lycaenidae	Lycaeninae	*Arhopala herculina*	-	X
Lycaenidae	Lycaeninae	*Caleta mindarus*	X	-
Lycaenidae	Lycaeninae	*Candalides margarita*	-	X
Lycaenidae	Lycaeninae	*Danis danis*	X	X
Lycaenidae	Lycaeninae	*Danis glaucopis*	X	-
Lycaenidae	Lycaeninae	*Danis* sp. (probably *melimnos*)	X	-
Lycaenidae	Lycaeninae	*Danis* sp. (probably *regalis*)	X	-
Lycaenidae	Lycaeninae	*Everes lacturnus*	-	X
Lycaenidae	Lycaeninae	*Hypochrysops chrysargyrus*	-	X
Lycaenidae	Lycaeninae	*Hypochrysops pythias*	-	X
Lycaenidae	Lycaeninae	*Hypolycaena phorbas*	-	X
Lycaenidae	Lycaeninae	*Jamides celeno*	X	-
Lycaenidae	Lycaeninae	*Perpheres perpheres*	X	X
Lycaenidae	Lycaeninae	*Pithecops dionisius*	X	X

continued

Family	Subfamily	Species	Forest	Garden/ Forest Edge
Lycaenidae	Lycaeninae	*Psychonotis caelius*	-	X
Nymphalidae	Danainae	*Danaus affinis*	X	-
Nymphalidae	Danainae	*Danaus philene*	-	X
Nymphalidae	Danainae	*Euploea boisduvalii*	-	X
Nymphalidae	Danainae	*Euploea nemertes*	X	-
Nymphalidae	Danainae	*Euploea* sp.	-	X
Nymphalidae	Ithoniinae	*Tellervo assarica*	X	X
Nymphalidae	Morphinae	*Hyantis hodeva*	X	X
Nymphalidae	Morphinae	*Taenaris artemis*	X	X
Nymphalidae	Morphinae	*Taenaris catops*	X	X
Nymphalidae	Morphinae	*Taenaris dimona*	-	X
Nymphalidae	Satyrinae	*Elymnias paradoxa*	X	-
Nymphalidae	Satyrinae	*Elymnias cybele*	X	-
Nymphalidae	Satyrinae	*Hypocysta isis*	X	-
Nymphalidae	Satyrinae	*Mycalesis duponchelii*	X	X
Nymphalidae	Satyrinae	*Mycalesis durga*	-	X
Nymphalidae	Satyrinae	*Mycalesis elia*	-	X
Nymphalidae	Satyrinae	*Mycalesis mehadeva*	-	X
Nymphalidae	Satyrinae	*Mycalesis mucia*	-	X
Nymphalidae	Satyrinae	*Mycalesis terminus*	X	X
Nymphalidae	Satyrinae	*Ypthima arctoa*	-	X
Nymphalidae	Nymphalinae	*Cethosia chrysippe*	X	X
Nymphalidae	Nymphalinae	*Cirrochroa regina*	-	X
Nymphalidae	Nymphalinae	*Doleschallia noorna*	-	X
Nymphalidae	Nymphalinae	*Hypolimnas deois*	X	X
Nymphalidae	Nymphalinae	*Hypolimnas bolina*	X	-
Nymphalidae	Nymphalinae	*Lexias aeropa*	X	X
Nymphalidae	Nymphalinae	*Neptis nausicaa*	X	X
Nymphalidae	Nymphalinae	*Pantoporia consimilis*	-	X
Nymphalidae	Nymphalinae	*Parthenos aspila*	-	X
Nymphalidae	Nymphalinae	*Precis hedonia*	X	X
Nymphalidae	Nymphalinae	*Precis Xellida*	-	X
Nymphalidae	Nymphalinae	*Terinos tethys*	X	-
Nymphalidae	Nymphalinae	*Xindula arsinoe*	X	X
Nymphalidae	Nymphalinae	*Yoma algina*	-	X
Nymphalidae	Apaturinae	*Apaturina erminea*	X	X
Nymphalidae	Apaturinae	*Cyrestis acilia*	X	X
Nymphalidae	Charaxinae	*Prothoe australis*	X	X
Nymphalidae	Charaxinae	*Polyura jupiter*	-	X

Appendix 17

List of butterflies recorded at Furu and Tiri Rivers, Mamberamo Basin, Papua, Indonesia

Henk van Mastrigt and Edy M. Rosariyanto

Species	Furu	Tiri
Papilionidae		
Atrophaneura polydorus	X	X
Graphium agamemnon	X	X
Graphium aristeus	X	X
Graphium eurypylus	X	X
Graphium sarpedon	X	X
Graphium wallacei	X	
Ornithoptera priamus	X	X
Papilio aegeus	X	X
Papilio ambrax	X	X
Papilio euchenor	X	X
Papilio ulysses	X	X
Troides oblongomaculatus		X
Pieridae		
Coliadinae		
Catopsilia pomona	X	X
Eurema blanda	X	X
Eurema candida	X	X
Eurema hecabe	X	X
Pierinae		
Appias ada	X	X
Appias celestina	X	X
Delias aruna		X
Delias ladas		X
Delias mysis		X
Elodina andropis		X
Saletara cycinna	X	X
Lycaenidae		
Riodininae		
Praetaxila statira	X	
Curetinae		
Curetis barsine	X	X

Species	Furu	Tiri
Lycaeninae		
Anthene lycaenoides	X	X
Arhopala helenita	X	X
Arhopala herculina	X	X
Arhopala madytus	X	X
Arhopala sp.	X	
Caleta mindarus	X	
Catochrysops panormus	X	X
Catochrysops sp.	X	
Danis danis	X	X
Danis melimnos	X	X
Danis sp. 1	X	X
Danis sp. 2	X	
Epimastidia inops	X	X
Erysichton lineata	X	X
Hypochlorosis ancharia	X	
Hypolycaena phorbas		X
Hypochrysops clarysargyrus		X
Hypochrysops polycletus	X	
Hypochrysops pythias	X	X
Ionolyce helicon	X	
Jamides aetherialis	X	
Jamides aleuas	X	
Jamides allectus	X	
Jamides coridus	X	X
Jamides sp.	X	
Nacaduba berenice		X
Nacaduba cyana	X	X
Nacaduba pactolus	X	X
Nacaduba sp. 1	X	
Nacaduba sp. 2	X	X
Nacaduba sp. 3	X	

continued

Species	Furu	Tiri
Nothodanis schaeffera	X	
Perpheres perpheres	X	X
Philiris agatha		X
Philiris fulgens	X	X
Philiris helena	X	X
Philiris moira	X	X
Philiris pagwi		X
Philiris sp. 1	X	X
Philiris sp. 2	X	
Philiris sp. 3	X	
Philiris sp. 4	X	
Philiris sp. 5	X	X
Pithecops dionisius	X	
Psychonotis caelius	X	X
Sahulana scintillata	X	X
Udara cardia	X	
Upolampes evena		X
Nymphalidae		
Libytheinae		
Libythea geoffroy		X
Ithomiinae		
Tellervo zoilus	X	
Danainae		
Euploea algea	X	X
Euploea phaenareta		X
Euploea sylvester		X
Euploea tulliolus		X
Euploea wallacei	X	
Ideopsis juventa	X	X
Parantica kirbyi		X
Tirumala hamata	X	X
Morphinae		
Hyantis hodeva	X	
Taenaris artemis	X	X
Taenaris catops	X	X
Taenaris cyclops	X	
Taenaris dimona	X	X
Satyrinae		
Elymnias cybele		X
Melanitis amabilis	X	X
Melanitis constantia	X	
Mycalesis aethiops	X	
Mycalesis asophis		X
Mycalesis duponchelli	X	

Species	Furu	Tiri
Mycalesis durga	X	X
Mycalesis elia		X
Mycalesis mehadeva	X	
Mycalesis mucia	X	X
Mycalesis phidon	X	X
Mycalesis shiva	X	X
Charaxinae		
Charaxes latona	X	
Polyura jupiter	X	X
Prothoe australis	X	X
Apaturinae		
Apaturina erminea	X	
Cyrestis achates	X	
Cyrestis acilia	X	X
Nymphalinae		
Doleschallia noorna	X	
Euthaliopsis aetion	X	
Hypolemas alimena	X	
Hypolemas antilope	X	
Hypolemas bolina		X
Hypolemas deois	X	X
Lexias aeropa	X	
Mynes geoffroyi	X	
Neptis nausicaa	X	X
Parthenos aspila	X	X
Pantoporia consimilis	X	X
Pantoporia venilia	X	
Phaedyma shepherdi	X	X
Precis hedonia	X	X
Precis villida	X	X
Yoma algina	X	X
Heliconiinae		
Algia felderi	X	
Cethosia cydippe	X	X
Cirrochroa regina	X	X
Cupha propose	X	X
Phalanta alcippe	X	X
Vagrans egista	X	X
Vindula arsinoe	X	X
Total	**109**	**89**

Appendix 18

Diversity of moths collected in the Dabra area, Papua, Indonesia

Henk van Mastrigt and Edy M. Rosariyanto

Superfamily	Family	Number of Genera	Identified Species	Unidentified Species	Total Number of Species
Large Moths					
Cossoidea	Cossidae	0	0	0	0
Zygaenoidea	Limacodidae	> 1	0	11	11
	Zygaenidae	> 1	0	3	3
Sphingoidea	Sphingidae	4	5	0	5
Bombycoidea	Bombycidae	1	0	1	1
	Saturniidae	0	0	0	0
	Lasiocampidae	0	0	0	0
Notodontoidea	Notodontidae	> 4	1	13	14
Noctuoidea	Lymantriidae	> 7	0	16	16
	Arctiidae	> 13	13	26	39
	Aganaidae	1	1	0	1
	Castniidae	0	0	0	0
Geometroidea	Callidulidae	> 2	1	2	3
	Drepanidae	> 3	1	8	9
	Uraniidae	> 5	5	2	7
	Geometridae	> 22	6	120	126
	Noctuidae	> 15	5	37	42
	Total	**> 79**	**38**	**239**	**277**
Skippers					
Hesperioidea	Hesperiidae	> 9	4	10	14
Small Moths					
Hepialoidea	Hepialidae	0	0	0	0
Tineoidea	Tineidae	1	0	1	1
Yponomeutoidea	Yponomeutidae	1		1	1
Gelechioidea	Oecophoridae	> 1		5	5
	Gelechiidae	1		1	1
	Lecithoceridae	> 1		3	3

continued

Superfamily	Family	Number of Genera	Identified Species	Unidentified Species	Total Number of Species
Tortricoidea	Tortricidae	> 2		4	4
	Psychidae	1		2	2
Sesioidea	Sesiidae	1	1		1
Choreutoidea	Choreutidae				0
Alucitoidea	Alucitidae				0
Hyblaeoidea	Hyblaeidae	1	0	1	1
Thyridoidea	Thryrididae	> 3	1	5	6
Pyraloidea	Pyralidae	> 8	10	106	116
	Total	>21	12	129	141
Grand Total		109	54	378	432

Appendix 19

Summary of fish collection/observation sites in the Yongsu and Dabra areas, Papua, Indonesia

Gerald R. Allen

YONGSU AREA

Site 1: Omu River, about 1 km S of Yemang Training Camp, 2°26.447'S, 140°29.090'E; elevation about 10 m, approximately 50 m upstream from sea; a small, clear, creek (2–4 m wide with pools to 0.7 m deep) with alternating riffles, pools, and small cascades; moderately fast flowing over sand, gravel, rocks, and boulders; through closed-canopy forest; water temperature 24.5°C; hand nets and underwater observations with mask and snorkel, G. Allen and party, 23 August 2000.

Site 2: Nantuke River, about 2 km SE of Yemang Training Camp, 2°27.103'S, 140°29.174'E; elevation about 70 m, approximately 2 km upstream from sea; clear pool (10 m in diameter to 3 m deep) at base of 15 m high waterfall; moderately fast flowing over gravel, cobbles, rocks, and boulders in open canopy forest; water temperature 25.6°C; hand nets and underwater observations with mask and snorkel, G. Allen and party, 24 August 2000.

Site 3: Junction of Nantuke River and Danyamo River, about 1.4 km SE of Yemang Training Camp, 2°26.761'S, 140°29.580'E; elevation about 5 m, approximately 200 m upstream from sea; small, clear creek (about 3–6 m wide with pools to 0.5 m deep) with alternating riffles, pools, and small cascades; moderately fast flowing over gravel, cobbles, and bedrock through open-canopy second growth forest; water temperature 24.6°C in Nantuke River and 26.6°C in Danyamo River; hand nets and underwater observations with mask and snorkel, G. Allen and party, 24 August 2000.

Site 4: Danyamo River, about 2.7 km SE of Yemang Training Camp, 2°27.411'S, 140°29.485'E; elevation about 70 m, approximately 2–3 km upstream from sea; large, clear creek (about 10–15 m wide with pools to 0.5 m deep); moderately fast flowing over sand, gravel, cobbles, and boulders through open canopy forest; water temperature 24.9°C; hand nets and underwater observation with mask and snorkel, also 0.5 kg rotenone, G. Allen and party, 25 August 2000.

Site 5: Sapari River, about 4–5 km NW of Yemang Training Camp, 2°26.262'S, 140°27.539'E; elevation about 15 m, approximately 1 km upstream from sea; large, clear creek (about 10 m wide with pools to 1.0 m deep) with alternating pools and cascades; moderately fast flowing over sand, gravel, cobbles, and boulders through open canopy forest; water temperature 24.7°C; hand nets and underwater observation with mask and snorkel, G. Allen and party, 26 August 2000.

Site 6: Danyamo-Nantuki River, about 1.4 km SE of Yemang Training Camp, 2°26.761'S, 140°29.580'E; elevation about 3 m, approximately 50 m upstream from sea; large, clear creek (about 10 m wide with pools to 1.0 m deep); moderately fast to slow flowing over sand, gravel, cobbles, and boulders through open canopy second growth forest; water temperature 25.5°C; hand nets and underwater observation with mask and snorkel, also small-meshed seine, G. Allen and party, 26 August 2000.

Site 7: Brackish pool next to Yemang Camp, 2°26.103'S, 140°29.082'E; elevation about 2 m, 5–20 m upstream from sea; large, tannin-stained, slightly turbid pool (about 35 m long and 5 m wide with depths to 1.3 m) perched at high tide mark and mainly fresh except at maximum high water; no current except slight flow at inlet, bottom consisting of beach sand and log debris, pool mainly shaded by secondary growth; water temperature 28.5°C; 1.0 kg rotenone, G. Allen and party, 28 August 2000.

Site 8: Yongsu Bay, adjacent to training camp, 2°26.103'S, 140°29.082'E, observations of marine fishes while using mask and snorkel at surface and free-diving to 5 m depth, G. Allen, 19–29 August 2000.

DABRA AREA

Site 1: Furu River, canoe landing area about 500 m upstream from junction with Idenburg River, approximately 3 km E of Dabra, 3°16.590'S, 138°38.295'E; small, slightly turbid stream (average 4 m wide and 4 cm deep with pools to 1.6 m), moderate to slow flow through open canopy second-growth forest; water temperature 25.6°C; underwater observations with mask and snorkel and 1 kg rotenone over 300 m section of stream; G. Allen, P. Boli, and O. Foisa, 2 September 2000.

Site 2: Furu River, about 100 m downstream from Furu Camp, approximately 4 km SE of Dabra; 3°17.091'S, 138°38.102'E; small, clear pool (4–5 m diameter and depth to 1.8 m) at base of 1 m-high cascade; moderately fast flowing over rock and sand bottom with log debris through closed-canopy rainforest; water temperature 24.9°C; underwater observations with mask and snorkel, also hand nets and small seine; G. Allen, P. Boli, and O. Foisa, 4 September 2000.

Site 3: Buare Lagoon, approximately 8 km E of Dabra, 3°15.741'S, 138°41.198'E; small, turbid channel (1.5–5.0 m wide to 1 m deep) draining oxbow lake; slow flowing over soft mud bottom through open-scrub forest; water temperature 33.1°C; 1 kg rotenone over 100 m section of stream; G. Allen, P. Boli, and O. Foisa, 5 September 2000.

Site 4: Esai River, approximately 4 km E of Dabra; 3°16.829'S, 138°38.327'E; small, clear stream (average 2 m wide and .4 cm deep) with partly submerged grass along bank; slow flowing over mud bottom through open canopy second growth forest; water temperature 26.1°C; small seine net; G. Allen, P. Boli, and O. Foisa, 5 September 2000.

Site 5: Tributary of Tiri River; approximately 4 km W of Dabra; 3°16.078'S, 138°34.832'E; small, turbid stream (average 4 m wide and 0.5 m deep with pools to 1.5 m), slow flowing over mud, gravel, and leaf litter bottom with occasional log jams,

through closed canopy rainforest; water temperature 24.6°C; 1 kg rotenone over 200 m section of stream; G. Allen, P. Boli, and O. Foisa, 6 September 2000.

Site 6: Side branch of Tiri River, about 200 m upstream from Tiri Camp, approximately 5 km SW of Dabra; 3°17.595'S, 138°34.839'E; small, slightly turbid overflow channel (3–5 m wide with pools to 1.8 m deep), moderate to slow flow over mainly sand and gravel bottom with occasional log jams, through closed canopy rainforest; water temperature 24.6°C; underwater observations with mask and snorkel and netting (hand nets and 15 m length seine); G. Allen, P. Boli, and O. Foisa, 7 September 2000.

Site 7: Side branch of Tiri River, about 200 m downstream from camp, approximately 5 km SW of Dabra, 3°17.595'S, 13 8°34.839'E; overflow channel forming broad, sandy corridor through partly closed canopy rainforest; many semi-stagnant pools to 1.3 m deep; no flow; most fishes from clear, circular pool about 4 m in diameter and 1.3 m deep; water temperature 24.2°C; small seine net; G. Allen and P. Boli, 8 September 2000.

Site 8: Dabra market; catch from local fishermen; fishes observed and photographed by G. Allen, 2–9 September 2000.

Appendix 20

Summary of freshwater fishes collected during the Yongsu training course, Papua, Indonesia

Gerald R. Allen

Abbreviations are as follows:
R = 3 or fewer individuals per site
O = 5–30 per site
C = many per site, often more than 100
Numbers for site 7 indicate the actual number of specimens collected with rotenone. Sites are described in Appendix 19.

Family/Species	1	2	3	4	5	6	7
Anguillidae							
Anguilla marmorata		R		R			1
Synganthidae							
Microphis bracyurus							3
Microphis leiaspis			R	R			
Tetrarogidae							
Tetraroge barbatus						R	
Ambassidae							
Ambassis miops						C	32
Terapontidae							
Mesopristes argenteus			O			O	3
Mesopristes cancellatus			O			R	
Kuhliidae							
Kuhlia marginata	R	C	C	C	C	C	
Kuhlia rupestris		O					1
Apogonidae							
Apogon hyalosoma							1
Sillaginidae							
Sillago sihama						R	
Carangidae							
Caranx sexfasciatus						R	

continued

Family/Species	Site number						
	1	2	3	4	5	6	7
Lutjanidae							
Lutjanus argentimaculatus						R	
Lutjanus fuscescens		R				R	
Mugilidae							
Crenimugil heterocheilus						C	
Liza subviridis							33
Eleotridae							
Belonobranchus belonobranchus		O	O	O	O	O	
Eleotris fusca				O		O	9
Ophiocara porocephala							2
Gobiidae							
Awaous sp.			O	O		O	
Glossogobius sp.							1
Lentipes multiradiatus				R			
Mugilogobius rambiae							1
Schismatogobius marmoratus		R		R			
Sicyopterus lagocephalus		O	C	C	C	C	
Sicyopterus longifilis		O	C	C	C	O	
Sicyopus mystax		O					
Stiphodon birdsong		O		O	O		
Stiphodon rutilaureus		C	O				
Stiphodon semoni	C		C	C			
Stenogobius beauforti							5
Scatophagidae							
Scatophagus argus							3
Siganidae							
Siganus vermiculatus							1

Appendix 21

Annotated checklist of fishes of the Yongsu area, Papua, Indonesia

Gerald R. Allen

The phylogenetic sequence of families appearing in this list follows the system used by major Australian museums and approximates that proposed in Nelson's *Fishes of the World* (2nd edition, 1984, John Wiley and Sons). Genera and species are arranged alphabetically within each family.

Text for each species includes a series of annotations, each separated by a semicolon. These annotations pertain to general habitat, detailed habitat, known altitudinal range, general activity mode, social behavior, major feeding type, food items, reproductive mode, maximum size, general distributional range, and additional comments pertinent to the present survey. The length is given as standard length (SL) for most species, which is the distance from the tip of the snout to the base of the caudal fin. Total length (TL) is given for a few fishes that do not have a clearly defined caudal fin (eels and plotosid catfishes for example).

ANGUILLIDAE—FRESHWATER EELS

Anguilla marmorata Quoy and Gaimard, 1824— Giant Long-Finned Eel
Creeks and rivers; cryptic; solitary; carnivore; fishes, crustaceans; spawns pelagic eggs; at least 90 cm TL; Indo-west Pacific; on high islands from East Africa to Marquesas.

SYNGNATHIDAE—PIPEFISHES AND SEAHORSES

Microphis brachyurus (Bleeker, 1853)—Short-tailed Pipefish
Mangrove estuaries, tidal creeks, and lowland streams; below about 10 m elevation; diurnal benthic; solitary; carnivore; tiny crustaceans; male broods eggs in pouch or on ventral surface; 21 cm SL; Indo-west Pacific.

Microphis leiaspis (Bleeker, 1853)—Barhead Pipefish
Creeks and rivers; usually below 100 m elevation; diurnal benthic; solitary; carnivore; tiny crustaceans; male broods eggs in pouch or on ventral surface; 18 cm SL; Indo-west Pacific.

WASPFISHES—TETRAROGIDAE

Tetraroge barbata (Cuvier, 1829)—Freshwater Waspfish
Mangrove estuaries, tidal creeks, and lowland streams; diurnal benthic; solitary; carnivore; fishes, crustaceans; spawns pelagic eggs; 8 cm SL; Indonesia, Philippines, and New Guinea.

CHANDIDAE—GLASSFISHES

Ambassis miops Gunther, 1871—Flag-tailed Perchlet
Creeks and rivers; usually below 100 m elevation; hovers in midwater; forms aggregations; omnivore; insects, microcrustaceans, fishes, algae; eggs spawned on weed or floating debris; 6.5 cm SL; Western Pacific.

TERAPONTIDAE—GRUNTERS

Mesopristes argenteus (Cuvier, 1829)—Silver Grunter
Mangrove estuaries, tidal creeks, and lowland streams; 0–4 m; diurnal benthic; solitary or in groups; carnivore; crustaceans, fishes; spawns pelagic eggs; 28 cm SL; Coastal streams of Indo-Australian Archipelago including Philippines, Indonesia, and New Guinea to Vanuatu.

Mesopristes cancellatus (Cuvier, 1829)—Tapiroid Grunter
Creeks and rivers; to an elevation of at least 200 m, often in hilly terrain; diurnal benthic; solitary or in groups; carnivore; crustaceans, fishes; spawns pelagic eggs; 23 cm SL; Western Pacific.

continued

KUHLIIDAE—FLAGTAILS

***Kuhlia marginata* (Cuvier, 1829)—Spotted Flagtail**
Creeks and rivers; 0–5 m; diurnal midwater; forms aggregations; carnivore; insects and larvae, crustaceans, fishes; spawns pelagic eggs; 18 cm SL; Western Pacific.

***Kuhlia rupestris* (Lacepede, 1802)—Jungle Perch**
Creeks and rivers; to about 100 m elevation; diurnal midwater; forms aggregations; carnivore; insects and larvae, crustaceans, fishes; spawns pelagic eggs; 30 cm SL; Indo-west Pacific.

APOGONIDAE—CARDINALFISHES

***Apogon hyalosoma* Bleeker, 1852—Mangrove Cardinalfish**
Mangrove estuaries, tidal creeks, and lowland streams; diurnal benthic; forms aggregations; carnivore; crabs, shrimps, worms, fishes; male broods eggs in mouth; 15 cm SL; Indo-west Pacific.

SILLAGINIDAE—WHITINGS

***Sillago sihama* (Forsskal, 1775)—Beach Whiting**
Mangrove estuaries, sandy beaches, and freshwater streams; 0–12 m; diurnal benthic; forms aggregations; carnivore; sand-dwelling invertebrates; spawns pelagic eggs; 23 cm SL; Indo-west Pacific.

CARANGIDAE—TREVALLIES OR JACKS

***Caranx sexfasciatus* Quoy & Gaimard, 1825**
Mainly marine, but young often inhabit mangrove estuaries and lower reaches of freshwater streams; roving predator; solitary or in groups; carnivore; fishes, crabs, lobsters; spawns pelagic eggs; 85 cm SL; Indo-Pacific to the Americas.

LUTJANIDAE—SNAPPERS

***Lutjanus argentimaculatus* (Forsskal, 1775)**
Mainly marine, but young often inhabit mangrove estuaries and lower reaches of freshwater streams; diurnal benthic; solitary; carnivore; fishes, crustaceans; spawns pelagic eggs; 100 cm SL; Indo-west and central Pacific.

***Lutjanus fuscescens* (Valenciennes, 1830)—Papuan Spotted Bass**
Lowland creeks and rivers; to at least 100 m elevation; diurnal benthic; solitary; carnivore; fishes, crustaceans; spawns pelagic eggs; to at least 80 cm SL; Western Pacific including China, Philippines, Indonesia, and New Guinea.

SCATOPHAGIDAE—SCATS

***Scatophagus argus* (Bloch, 1788)—Spotted Scat**
Mangrove estuaries, coastal bays, tidal creeks, and lowland streams; below about 50 m elevation; forms benthic grazing schools; forms aggregations; omnivore; small benthic invertebrates, algae, detritus; spawns pelagic eggs; 30 cm SL; Indo-west Pacific from India to Society Islands.

MUGILIDAE—MULLETS

***Crenimugil heterocheilus* Bleeker, 1855—Fringe-lipped Mullet**
Lowland creeks and rivers; below about 50 m elevation; forms benthic grazing schools; forms aggregations; herbivore; bottom detritus and plants; spawns pelagic eggs; 50 cm SL; Coastal streams of Indo-Australian Archipelago.

***Liza subviridis* (Valenciennes, 1836)—Greenback Mullet**
Mangrove estuaries, lowland creeks, and rivers; to at least 100 m elevation; forms benthic grazing schools; forms aggregations; omnivore; algae, organic detritus; spawns pelagic eggs; to 26 cm SL; Indo-west and central Pacific from Persian Gulf to Polynesia.

GOBIIDAE—GOBIES

***Awaous* sp.—Roman-nosed Goby**
Creeks and rivers; usually below 100 m elevation; rests on bottom; solitary or in groups; omnivore; algae, small crustaceans; parental care of demersal eggs; about 10 cm SL; northern New Guinea.

***Glossogobius* sp.—False Celebes Goby**
Creeks and rivers; 0-5 m; rests on bottom; solitary; carnivore; crustaceans, small fishes; parental care of demersal eggs; 12 cm SL; Western Pacific; formerly confused with *G. celebius*, but apparently is undescribed.

***Lentipes multiradiatus*—Cyclops Cling-goby**
Creeks and rivers; diurnal benthic; solitary or in groups; omnivore; algae, micro-invertebrates; parental care of demersal eggs; to at least 4.5 cm SL; known only from a single specimen collected during the current survey.

***Mugilogbius rambaiae* (Smith, 1945)—Shoulder-spot Goby**
Mangrove estuaries, tidal creeks, and lowland streams; below about 10 m elevation; rests on bottom; solitary or in groups; omnivore; algae, benthic invertebrates; parental care of demersal eggs; 3 cm SL; Indo-Australian Archipelago.

continued

***Schismatogobius marmoratus* (Peters, 1868)—Scaleless Goby**
Lowland creeks and rivers; usually below 100 m elevation; diurnal benthic; solitary; carnivore; small invertebrates; parental care of demersal eggs; 2.5 cm SL; Western Pacific including Japan, Philippines, Indonesia, and New Guinea.

***Sicyopterus lagocephalus* (Pallas, 1770)—Rabbithead Cling-goby**
Creeks and rivers; usually below 500 m elevation; diurnal benthic; solitary or in groups; herbivore; filamentous algae growing on rock surfaces; parental care of demersal eggs; to at least 80 cm SL; Indo-west Pacific on high islands.

***Sicyopterus longifilis* de Beaufort, 1912—Threadfin Goby**
Creeks and rivers; diurnal benthic; solitary or in groups; herbivore; filamentous algae growing on rock surfaces; parental care of demersal eggs; 75 cm SL; Coastal streams of Indo-Australian Archipelago including Philippines, Indonesia, and New Guinea.

***Sicyopus mystax* Watson and Allen, 1999—Moustached Cling-goby**
Swift coastal streams; to at least 200 m elevation; diurnal benthic; solitary or in groups; herbivore; algae; parental care of demersal eggs; 4 cm SL; northern New Guinea in hilly terrain.

***Stenogobius beauforti* (Weber, 1908)—Beaufort's Goby**
Lowland creeks and rivers; rests on bottom; solitary or in groups; herbivore; algae; parental care of demersal eggs; 5 cm SL; northern New Guinea.

***Stiphodon birdsong* Watson, 1996—Birdsong's Cling-goby**
Creeks and rivers; usually in hilly terrain to about 400 m elevation; rests on bottom; solitary or in groups; herbivore; grazes algae from rocky surfaces; parental care of demersal eggs; 2.3 cm SL; northern New Guinea, also Halmehera Island, Indonesia.

***Stiphodon rutilaureus* Watson, 1996—Red and Gold Cling-goby**
Creeks and rivers; to about 400 m elevation; rests on bottom; solitary or in groups; herbivore; grazes algae from rocky surfaces; parental care of demersal eggs; 3 cm SL; Indo-Australian Archipelago from Waigeo and Batanta (Raja Ampat Islands, Papua) to Vanuatu.

***Stiphodon semoni* Weber, 1895—Neon Goby**
Creeks and rivers; usually coastal streams, but as far as 120 km inland; rests on bottom; solitary or in groups; herbivore; grazes algae from rocky surfaces; parental care of demersal eggs; 3.5 cm SL; Indo-Australian Archipelago including Philippines, Indonesia, and New Guinea.

ELEOTRIDAE—GUDGEONS

***Belobranchus belobranchus* (Valenciennes, 1837)—Throatspine Gudgeon**
Creeks and rivers; 0-5 m; rests on bottom; solitary or in groups; carnivore; fishes, small crustaceans; parental care of demersal eggs; 16 cm SL; Coastal streams of Indo-Australian Archipelago including Philippines, Indonesia, and New Guinea.

***Eleotris fusca* (Bloch and Schneider, 1801)—Brown Gudgeon**
Mangrove estuaries, tidal creeks, and lowland streams; rests on bottom; solitary or in groups; carnivore; insects, crustaceans, small fishes; parental care of demersal eggs; 15 cm SL; Indo-west Pacific from East Africa to high volcanic islands of the Pacific.

***Ophiocara porocephala* (Valenciennes, 1837)—Spangled Gudgeon**
Mangrove estuaries, tidal creeks, and lowland streams; below about 20 m elevation; hovers in midwater; solitary or in groups; carnivore; benthic invertebrates; parental care of demersal eggs; 20 cm SL; Indo-west Pacific.

SIGANIDAE—SPINEFEET OR RABBITFISHES

***Siganus vermiculatus* (Valenciennes, 1835)—Vermiculated Spinefoot**
Brackish mangrove estuaries and lower reaches of freshwater streams; diurnal benthic; solitary or in groups; herbivore; algae and seagrasses; spawns pelagic eggs; 35 cm SL; Indo-west and central Pacific.

Appendix 22

List of shallow coral reef fishes of Yongsu Bay, Papua, Indonesia

Gerald R. Allen

Fishes recorded by G. Allen while snorkelling (approximately 6 hours between 19–28 August 2000) in Yongsu Bay, Papua.

Carcharhinidae
Carcharhinus
   melanopterus (Quoy & Gaimard, 1824)

Hemigaleidae
Triaenodon
   obesus (Ruppell, 1835)

Megalopidae
Megalops
   cyprinoides (Broussonet, 1782)

Clupeidae
Spratelloides
   gracilis (Temminck & Schlegel, 1846)

Synodontidae
Synodus
   dermatogenys Fowler, 1912

Holocentridae
Myripristis
   berndti Jordan & Evermann, 1902
   kuntee Valenciennes, 1831
   violacea Bleeker, 1851
Neoniphon
   sammara (Forsskal, 1775)
Sargocentron
   cornutum (Bleeker, 1853)
   microstomus (Gunther, 1859)

Aulostomidae
Aulostomus
   chinensis (Linnaeus, 1766)

Scorpaenidae
Sebastapistes
   cyanostigma (Bleeker, 1856)

Serranidae
Cephalopholis
   argus Bloch & Schneider, 1801
   leopardus (Lacepede, 1802)
   urodeta (Schneider, 1801)

Epinephelus
   hexagonatus (Bloch & Schneider, 1801)
Variola
   louti (Forsskal, 1775)

Cirrhitidae
Cirrhitichthys
   oxycephalus (Bleeker, 1855)
Cirrhitus
   pinnulatus (Schneider, 1801)
Paracirrhites
   arcatus (Cuvier, 1829)
   forsteri (Schneider, 1801)

Apogonidae
Apogon
   taeniophorus Regan, 1908

Malacanthidae
Malacanthus
   latovittatus (Lacepede, 1798)

Echeneidae
Echeneis
   naucrates Linnaeus, 1758

Carangidae
Caranx
   melampygus Cuvier, 1833
   papuensis Alleyne and Macleay, 1877
Scomberoides
   tala (Cuvier, 1832)

Lutjanidae
Aphareus
   furca (Lacepede, 1802)
Lutjanus
   bohar (Forskal, 1775)
   fulviflamma (Forskal, 1775)
   gibbus (Forskal, 1775)
   rivulatus (Cuvier, 1828)
   semicinctus Quoy and Gaimard, 1824
Macolor
   macularis Fowler, 1931
   niger (Forsskal, 1775)

continued

Caesionidae
Caesio
   caerulaurea Lacepede, 1802
   lunaris Cuvier, 1830
Pterocaesio
   tile (Cuvier, 1830)

Nemipteridae
Scolopsis
   bilineatus (Bloch, 1793)

Haemulidae
Diagramma
   pictum (Thunberg, 1792)
Plectorhinchus
   gibbosus (Lacepede, 1802)
   orientalis (Bloch, 1793)

Lethrinidae
Lethrinus
   atkinsoni Seale, 1909
   harak (Forskkal, 1775)
Monotaxis
   grandoculis (Forsskal, 1775)

Mullidae
Mulloidichthys
   flavolineatus (Lacepede, 1802)
Parupeneus
   barberinus (Lacepede, 1801)
   bifasciatus (Lacepede, 1801)
   cyclostomus (Lacepede, 1802)
Upeneus
   tragula Richardson, 1846

Pempheridae
Pempheris
   oualensis Cuvier, 1831
   vanicolensis Cuvier, 1831

Kyphosidae
Kyphosus
   cinerascens (Forskal, 1775)
   vaigiensis (Quoy & Gaimard, 1825)

Ephippidae
Platax
   orbicularis (Forskal, 1775)
   teira (Forsskal, 1775)

Chaetodontidae
Chaetodon
   auriga Forskal, 1775
   baronessa Cuvier, 1831
   citrinellus Cuvier, 1831
   ephippium Cuvier, 1831
   kleinii Bloch, 1790
   lineolatus Cuvier, 1831
   lunulatus Quoy and Gaimard, 1824
   meyeri Schneider, 1801
   ornatissimus Cuvier, 1831
   oxycephalus Bleeker, 1853
   rafflesi Bennett, 1830
   semeion Bleeker, 1855

   trifascialis Quoy & Gaimard, 1824
   unimaculatus Bloch, 1787
   vagabundus Linnaeus, 1758
Forcipiger
   flavissimus Jordan & McGregor, 1898
Heniochus
   chrysostomus Cuvier, 1831
   varius (Cuvier, 1829)

Pomacanthidae
Centropyge
   bicolor (Bloch, 1798)
   vroliki (Bleeker, 1853)
Pygoplites
   diacanthus (Boddaert, 1772)

Pomacentridae
Abudefduf
   notatus (Day, 1869)
   septemfasciatus (Cuvier, 1830)
   sexfasciatus Lacepede, 1802
   sordidus Forskal, 1775
   vaigiensis (Quoy & Gaimard, 1825)
Amphiprion
   clarkii (Bennett, 1830)
   melanopus Bleeker, 1852
   percula (Lacepede, 1802)
Chromis
   atripectoralis Welander & Schultz, 1951
   lepidolepis Bleeker, 1877
   lineata Fowler & Bean, 1928
   margaritifer Fowler, 1946
   ternatensis (Bleeker, 1856)
   viridis (Cuvier, 1830)
   weberi Fowler & Bean, 1928
   xanthura (Bleeker, 1854)
Chrysiptera
   brownriggii (Bennett, 1828)
   rex (Snyder, 1909)
   unimaculata (Cuvier, 1830)
Dascyllus
   reticulatus (Richardson, 1846)
   trimaculatus (Ruppell, 1928)
Neoglyphidodon
   crossi Allen, 1991
   melas (Cuvier, 1830)
Neopomacentrus
   azysron (Bleeker, 1877)
Plectroglyphidodon
   dickii (Lienard, 1839)
   lacrymatus (Quoy & Gaimard, 1824)
   leucozonus (Bleeker, 1859)
   phoenixensis (Schultz, 1943)
Pomacentrus
   bankanensis Bleeker, 1853
   brachialis Cuvier, 1830
   coelestis Jordan & Starks, 1901
   lepidogenys Fowler & Bean, 1928
   moluccensis Bleeker, 1853

Stegastes
   albifasciatus (Schlegel & Muller, 1839)
   fasciolatus (Ogilby, 1889)

Mugilidae
Crenimugil
   crenilabis (Forsskal, 1775)

Labridae
Anampses
   caeruleopunctatus Ruppell, 1828
   melanurus Bleeker, 1857
Bodianus
   diana (Lacepede, 1802)
   mesothorax Schneider, 1801
Cheilinus
   trilobatus Lacepede, 1802
   undulatus Ruppell, 1835
Coris
   aygula Lacepede, 1802
   gaimardi (Quoy & Gaimard, 1824)
Epibulus
   insidiator (Pallas, 1770)
Gomphosus
   varius Lacepede, 1801
Halichoeres
   argus (Bloch and Schneider, 1801)
   hortulanus (Lacepede, 1802)
   margaritaceus (Valenciennes, 1839)
   marginatus Ruppell, 1835)
   miniatus (Valenciennes, 1839
   scapularis Bennett, 1832
Hemigymnus
   fasciatus Bloch, 1792
   melapterus Bloch, 1791
Labrichthys
   unilineatus (Guichenot, 1847)
Labroides
   dimidiatus (Valenciennes, 1839)
   pectoralis Randall and Springer, 1975
Macropharyngodon
   meleagris (Valenciennes, 1839)
Oxycheilinus
   diagrammus (Lacepede, 1802)
Pseudodax
   moluccanus (Valenciennes, 1840)
Stethojulis
   bandanensis (Bleeker, 1851)
   trilineata (Bloch and Schneider, 1801)
Thalassoma
   amblycephalum (Bleeker, 1856)
   hardwicke (Bennett, 1828)
   jansenii Bleeker, 1856
   purpureum (Forsskal, 1775)
   quinquevittatum (Lay & Bennett, 1839)

Scaridae
Chlorurus
   bleekeri (de Beaufort, 1940)
   sordidus (Forsskal, 1775)
Scarus
   frenatus Lacepede, 1802
   niger Forsskal, 1775

   oviceps Valenciennes, 1839
   pyrrhurus (Jordan and Seale, 1906)
   quoyi Valenciennes, 1840
   rivulatus Valenciennes, 1840
   rubroviolaceus Bleeker, 1849
   spinus (Kner, 1868)

Pinguipedidae
Parapercis
   millepunctata (Gunther, 1860)

Blenniidae
Blenniella
   chrysospilos (Bleeker, 1857)
   lineatus (Valenciennes, 1836)
Cirripectes
   castaneus Valenciennes, 1836
   polyzona (Bleeker, 1868)
   stigmaticus Strasburg & Schultz, 1953
Entomacrodus
   striatus (Quoy & Gaimard, 1836)

Pholidichthyidae
Pholidichthys
   leucotaenia Bleeker, 1856

Tripterygiidae
Enneapterygius
   tutuilae Jordan & Seale, 1906

Gobiidae
Valenciennea
   sexguttata (Valenciennes, 1837)
   strigata (Broussonet, 1782)

Microdesmidae
Ptereleotris
   evides (Jordan & Hubbs, 1925)
   microlepis Bleeker, 1856
   zebra (Fowler, 1938)

Acanthuridae
Acanthurus
   blochi Valenciennes, 1835
   guttatus Forster and Schneider, 1801
   lineatus (Linnaeus, 1758)
   maculiceps (Ahl, 1923)
   nigricans (Linnaeus, 1758)
   nigricaudus Duncker and Mohr, 1929
   nigrofuscus (Forskal, 1775)
   olivaceus Bloch & Schneider, 1801
   pyroferus Kittlitz, 1834
   triostegus (Linnaeus, 1758)
Ctenochaetus
   striatus (Quoy & Gaimard, 1824)
Naso
   caeruleacauda Randall, 1994
   lituratus (Bloch & Schneider, 1801)
   unicornis Forskal, 1775
Zebrasoma
   scopas Cuvier, 1829
   veliferum Bloch, 1797

continued

Zanclidae
Zanclus
  cornutus Linnaeus, 1758

Siganidae
Siganus
  argenteus (Quoy & Gaimard, 1824)
  doliatus Cuvier, 1830
  puellus (Schlegel, 1852)
  spinus (Linnaeus, 1758)
  vulpinus (Schlegel & Muller, 1844)

Bothidae
Bothus
  mancus Broussonet, 1782

Balistidae
Balistapus
  undulatus (Park, 1797)
Balistoides
  viridescens (Bloch & Schneider, 1801)
Melichthys
  vidua (Solander, 1844)
Rhinecanthus
  rectangulus (Bloch and Schneider, 1801)
Sufflamen
  chrysoptera (Bloch & Schneider, 1801)

Monacanthidae
Aluterus
  scriptus (Osbeck, 1765)
Amanses
  scopas Cuvier, 1829
Cantherines
  fronticinctus (Gunther, 1866)
  pardalis Ruppell, 1866)
Oxymonacanthus
  longirostris Bloch & Schneider, 1801

Ostraciidae
Ostracion
  meleagris Shaw, 1796

Tetraodontidae
Arothron
  nigropunctatus (Bloch & Schneider, 1801)
Canthigaster
  amboinensis (Bleeker, 1865)

Appendix 23

Summary of fishes collected on the RAP survey in the Mamberamo River drainage, Papua, Indonesia

Gerald R. Allen

Abbreviations are as follows:

R = 3 or fewer individuals per site

O = 5–30 individuals per site

C = many per site, often more than 100

Sites are described in Appendix 19 and abbreviations are explained in Appendix 20.

Family/Species				Site number				
	1	2	3	4	5	6	7	8
Anguillidae								
Anguilla bicolor	R				R			
Ariidae								
Arius solidus								C
Arius utarus								C
Arius velutinus								C
Plotosidae								
Neosilurus novaeguinea	R							
Melanotaeniidae								
Chilatherina fasciata	C	C		C	C	C	C	
Glossolepis multisquamatus	C		R					
Melanotaenia praecox				R			C	
Melanotaenia vanheurni								
Terapontidae								
Hephaestus transmontanus	R					O		
Apogonidae								
Glossamia beauforti	O	O		O	O	O	O	
Eleotridae								
Giurus margaritaceus	O	O	O		C	O	C	
Mogurnda nesolepis	R			R	R		O	
Oxyeleotris fimbriata		O		R				
Oxyeleotris heterodon			C			R		

continued

Family/Species	Site number							
	1	2	3	4	5	6	7	8
Gobiidae								
Eugnathogobius tigrellus		O					R	
Glossogobius bulmeri	O	O				O		
Glossogobius koragensis	O		O					
Introduced fishes								
Cyprinidae								
Cyprinus carpio					O			C
Barbodes gonionotus			O		O			C
Puntius orphoides	O	O				O		
Clariidae								
Clarias batrachus						R		
Cichlidae								
Oreochromis mossambica	R		R					0

Appendix 24

Annotated checklist of fishes recorded from the Mamberamo River system, Papua, Indonesia

Gerald R. Allen

The phylogenetic sequence of families appearing in this list follows the system used by major Australian museums and approximates that proposed in Nelson's *Fishes of the World* (2nd edition, 1984, John Wiley and Sons). Genera and species are arranged alphabetically within each family.

Text for each species includes a series of annotations, each separated by a semicolon. These annotations pertain to general habitat, detailed habitat, known altitudinal range, general activity mode, social behavior, major feeding type, food items, reproductive mode, maximum size, general distributional range, and additional comments pertinent to the present survey. The length is given as standard length (SL) for most species, which is the distance from the tip of the snout to the base of the caudal fin. Total length (TL) is given for a few fishes that do not have a clearly defined caudal fin (eels and plotosid catfishes for example).

ANGUILLIDAE—FRESHWATER EELS

Anguilla bicolor McClelland, 1844—Indian Short-finned Eel
Creeks and rivers; to at least 1,000 m elevation; cryptic; solitary; carnivore; fishes, crustaceans; spawns pelagic eggs; 60 cm TL; Indo-west Pacific.

ARIIDAE—FORKTAIL CATFISHES

Arius solidus Herre, 1935—Hard-Palate Catfish
Large rivers; to about 100 m elevation; diurnal benthic; solitary or in groups; omnivore; insects, crustaceans, fishes, worms, plants; male broods eggs in mouth; 60 cm SL; Ramu, Sepik, and Mamberamo River systems of northern New Guinea.

Arius utarus Kailola, 1990—Northern Rivers Catfish
Large rivers; to at least 100 m elevation; diurnal benthic; solitary or in groups; omnivore; insects, prawns, and fishes; male broods eggs in mouth; 55 cm SL; Ramu, Sepik, and Mamberamo River systems of northern New Guinea.

Arius velutinus (Weber, 1909)—Papillate Catfish
Lowland creeks and rivers; diurnal benthic; solitary or in groups; omnivore; insects, higher plants; male broods eggs in mouth; 60 cm SL; Northern New Guinea; Ramu, Sepik, and Mamberamo River systems of northern New Guinea.

PLOTOSIDAE—EEL-TAILED CATFISH

Neosilurus idenburgi (Nichols, 1940)—Idenburg Tandan
Creeks and rivers; to at least 800 m elevation; nocturnal benthic; solitary or in groups; carnivore; insects, crustaceans, molluscs, worms, fishes; demersal eggs with no parental care; 27 cm TL; northern New Guinea.

Neosilurus novaeguineae (Weber, 1908)—New Guinea Tandan
Creeks and rivers; nocturnal benthic; solitary or in groups; carnivore; insects, crustaceans, molluscs, worms, fishes; demersal eggs with no parental care; 21 cm SL; Northern New Guinea from Ramu River, PNG westward to Wapoga River system, Papua.

HEMIRAMPHIDAE—HALFBEAKS OR GARFISHES

Zenarchopterus alleni Collette, 1982—Allen's River Garfish
Creeks and rivers; to about 100 m elevation; surface swimmer; forms aggregations; carnivore; floating insects; eggs spawned on weed or floating debris; to at least 13 cm SL; known on the basis of a single male specimen collected in 1920 at Batavia Bivak on the Mamberamo River.

Zenarchopterus kampeni (Weber, 1913)—Sepik River Garfish
Creeks and rivers; to at least 100 m elevation; surface swimmer; forms aggregations; carnivore; floating insects; eggs spawned on weed or floating debris; to 16.5 cm SL; Ramu, Sepik, and Mamberamo River systems of northern New Guinea.

continued

MELANOTAENIIDAE—RAINBOWFISHES

Chilatherina bleheri Allen, 1985—Bleher's Rainbowfish
Creeks and lakes; to 430 m elevation; diurnal mid-water; forms aggregations; carnivore; insects, aquatic insect larvae, microcrustaceans; eggs spawned on weed or floating debris; 12 cm SL; Known only from Danau Biru and its small inlet creeks, Mamberamo River system, Papua.

Chilatherina crassispinosa (Weber, 1913)
Creeks and rivers; between 100–600 m elevation; diurnal mid-water; forms aggregations; carnivore; insects, aquatic insect larvae, microcrustaceans; eggs spawned on weed or floating debris; 90 mm SL; northern New Guinea.

Chilatherina fasciata (Weber, 1913)—Barred Rainbowfish
Creeks, rivers, and lakes; to about 400–500 m elevation; diurnal midwater; forms aggregations; omnivore; insects, aquatic insect larvae, algae; eggs spawned on weed or floating debris; 10 cm SL; northern New Guinea.

Glossolepis multisquamatus (Weber and de Beaufort, 1922)—Sepik Rainbowfish
Creeks, rivers, and lakes; to at least 100 m elevation; diurnal mid-water; forms aggregations; carnivore; insects, aquatic insect larvae, microcrustaceans; eggs spawned on weed or floating debris; 10 cm SL; Ramu, Sepik, and Mamberamo River systems of northern New Guinea.

Melanotaenia maylandi Allen, 1982—Mayland's Rainbowfish
Small creeks; to at least 450 m elevation; diurnal mid-water; forms aggregations; omnivore; insects and their larvae, crustaceans, plants; eggs spawned on weed or floating debris; 9 cm SL; known only from a few small creeks near Danau Biru, Mamberamo River system, Papua.

Melanotaenia praecox (Weber and de Beaufort, 1922)—Dwarf Rainbowfish
Swamps and creeks; to elevation of about 100 m; diurnal mid-water; forms aggregations; omnivore; insects and their larvae, crustaceans, plants; eggs spawned on weed or floating debris; 5 cm SL; Mamberamo and Wapoga River systems, northern Papua.

Melanotaenia vanheurni (Weber and de Beaufort, 1922)—Van Heurn's Rainbowfish
Creeks and rivers; to at least 300 m elevation; diurnal mid-water; forms aggregations; omnivore; insects and their larvae, crustaceans, plants; eggs spawned on weed or floating debris; 16 cm SL; Mamberamo River system, Papua.

AMBASSIDAE—GLASSFISHES

Parambassis altipinnis Allen, 1982—High-finned Glass Perchlet
Creeks and rivers; to at least 100 m elevation; hovers in mid-water; forms aggregations; carnivore; insects, crustaceans, small fishes; eggs spawned on weed or floating debris; 12 cm SL; known only from 20 specimens collected in 1920 at Prauwenbivak, Mamberamo River, Papua.

Parambassis confinis (Weber, 1913)—Sepik Glass Perchlet
Creeks and rivers; to at least 300 m elevation; hovers in mid-water; forms aggregations; carnivore; insects, crustaceans, small fishes; eggs spawned on weed or floating debris; to 10 cm SL; Ramu, Sepik, and Mamberamo River systems of northern New Guinea.

TERAPONTIDAE—GRUNTERS

Hephaestus transmontanus (Mees and Kailola, 1977)—Sepik Grunter
Upland creeks and rivers; to at least 1,500 m elevation; roving predator; solitary or in groups; carnivore; insects, crustaceans, molluscs, fishes, frogs; demersal eggs with no parental care; 13 cm SL; Ramu, Sepik, Mamberamo, and Wapoga River systems of northern New Guinea.

APOGONIDAE—CARDINALFISHES

Glossamia beauforti (Weber, 1908)—Beaufort's Mouth Almighty
Creeks and rivers; lowlands to at least 400 m elevation; hovers in mid-water; solitary; carnivore; fishes and crustaceans; male broods eggs in mouth; 16 cm SL; Lake Sentani westward to Wapoga River system, northern Papua.

Glossamia gjellerupi (Weber & de Beaufort, 1929)—Gjellerup's Mouth Almighty
Creeks and rivers; to at least 200 m elevation; hovers in mid-water; solitary; carnivore; fishes, crustaceans; male broods eggs in mouth; to 16 cm SL; northern New Guinea from Markham River, PNG westward to Mamberamo River system, Papua.

MUGILIDAE—MULLETS

Mugilid sp.
Coastal and estuarine waters, also large rivers; diurnal mid-water; solitary or in groups; herbivore; demersal eggs with no parental care; Indo-west Pacific; local inhabitants informed the author that a species of mullet occurs in the river, but none were captured during the survey for positive identification; probably one of the three species of *Liza* reported from the Sepik-Mamberamo system (see Appendix 4).

continued

GOBIIDAE—GOBIES

***Glossogobius bulmeri* Whitley, 1959—Bulmer's Goby**
Upland creeks and rivers; to at least 1070 m elevation; rests on bottom; solitary; carnivore; crustaceans, small fishes; parental care of demersal eggs; 10 cm SL; northern New Guinea from Sepik River, PNG to Wapogo River system, Papua.

***Glossogobius koragensis* Herre, 1935—Sepik Goby**
Lowland creeks and rivers; rests on bottom; solitary; carnivore; crustaceans, small fishes; parental care of demersal eggs; 17 cm SL; Ramu, Sepik, Mamberamo, and Wapoga River systems of northern New Guinea.

***Eugnathogobius tigrellus* (Nichols, 1951)—Tiger Goby**
Lowland creeks to at least 100 m elevation; diurnal benthic; in pairs or small groups; parental care of demersal eggs; 3 cm SL; Mamberamo River system, Papua; known only from the vicinity of Dabra.

ELEOTRIDAE—GUDGEONS

***Giurus margaritaceus* (Valenciennes, 1837)—Snakehead Gudgeon**
Creeks, rivers, and lakes; to at least 600 m elevation; rests on bottom; solitary or in groups; omnivore; insects, crustaceans, plants; parental care of demersal eggs; 20 cm SL; Indo-west Pacific.

***Mogurnda aurofodinae* Whitley, 1938—Northern Mogurnda**
Swamps and creeks; to at least 1200 m elevation; hovers in mid-water; solitary; carnivore; insects, crustaceans, small fishes; parental care of demersal eggs; 10 cm SL; northern New Guinea.

***Mogurnda nesolepis* (Weber, 1908)—Yellowbelly Gudgeon**
Lowland creeks and rivers; usually below 100 m elevation; hovers in midwater; solitary; carnivore; insects, crustaceans, small fishes; parental care of demersal eggs; 3.5 cm TL; northern New Guinea from Markham River, PNG westward to Wapoga River system, Papua.

***Oxyeleotris fimbriata* (Weber, 1908)—Fimbriate Gudgeon**
Creeks, rivers, and lakes; 10–1,500 m elev.; rests on bottom; solitary; carnivore; insects, molluscs, crustaceans, fishes; parental care of demersal eggs; 16 cm SL; New Guinea and northern Australia; one of few purely freshwater fishes found on both sides of New Guinea's Central Dividing Range.

***Oxyeleotris heterodon* (Weber, 1908)—Sentani Gudgeon**
Creeks, rivers, and lakes; to at least 100 m elevation; rests on bottom; solitary; carnivore; insects, molluscs, crustaceans, fishes; parental care of demersal eggs; 41 cm SL; Ramu, Sepik, and Mamberamo River systems of northern New Guinea.

INTRODUCED SPECIES

CHANNIDAE—SNAKEHEADS

***Channa striata* (Bloch, 1793)—Striped Snakehead**
Swamps and creeks; to an elevation of at least 200 m; diurnal benthic; solitary; carnivore; fishes, crustaceans, frogs, snakes, insects; 90 cm SL; introduced from W. Indonesia.

CYPRINIDAE—CYPRINIDS

***Barbodes gonionotus* (Bleeker, 1850)—Java Barb**
Creeks, rivers, and lakes; diurnal benthic; solitary or in groups; omnivore; algae, detritus, crustaceans, insects, worms; demersal eggs with no parental care; 35 cm SL; introduced from SE Asia.

***Cyprinus carpio* Linnaeus, 1758—Common Carp**
Creeks, rivers, and lakes; diurnal benthic; solitary or in groups; omnivore; algae, detritus, crustaceans, insects, worms; demersal eggs with no parental care; 120 cm; widely introduced; native to Eurasia.

***Puntius orphoides* (Valenciennes, 1842)—Spot-tailed Barb**
Creeks and rivers; diurnal benthic; solitary or in groups; omnivore; algae, detritus, crustaceans, insects, worms; demersal eggs with no parental care; 25 cm SL; introduced from SE Asia.

CICHLIDAE—CICHLIDS

***Oreochromis mossambica* (Peters, 1852)—Tilapia**
Creeks, rivers, and lakes; diurnal benthic; solitary or in groups; herbivore; algae; parental care of demersal eggs; 30 cm SL; introduced from East Africa.

CLARIIDAE—WALKING CATFISHES

***Clarias batrachus* (Linnaeus, 1758)—Walking Catfish**
Creeks, rivers, and lakes; diurnal benthic; solitary; to at least 60 cm SL; introduced from SE Asia.

Appendix 25

List of fish species recorded to date from the Mamberamo, Sepik, and Ramu Rivers of northern New Guinea

Gerald R. Allen

Species restricted to this region (the Intermontane Trough) are indicated with an asterisk (*).

Family/Species	Mamberamo	Sepik	Ramu
Carcharhinidae			
Carcharhinus leucas			X
Pristidae			
Pristis microdon		X	X
Megalopidae			
Megalops cyprinoides		X	X
Anguillidae			
Anguilla bicolor	X	X	X
Anguilla marmorata		X	X
Clupeidae			
Escualosa throacata			X
Chanidae			
Chanos chanos		X	
Ariidae			
Arius coatesi*		X	X
Arius nox*		X	X
Arius solidus*	X	X	X
Arius utarus*	X	X	X
Arius velutinus*	X	X	X
Plotosidae			
Neosilurus gjellerupi*		X	X
Neosilurus idenburgi	X	X	X
Neosilurus novaeguinea	X	X	X
Neosilurus sp.*			X
Melanotaeniidae			
Chilatherina bleheri	X		
Chilatherina bulolo*			X
Chilatherina campsi		X	X
Chilatherina crassispinosa	X	X	X

Family/Species	Mamberamo	Sepik	Ramu
Chilatherina fasciata	X	X	X
Glossolepis maculosus*			X
Glossolepis multisquamatus*	X	X	X
Glossolepis ramuensis			X
Melanotaenia affinis		X	X
Melanotaenia maylandi*	X		
Melanotaenia praecox*	X		
Melanotaenia vanheurni*	X		
Hemiramphidae			
Zenarchopterus alleni*	X		
Zenarchopterus kampeni*	X	X	X
Syngnathidae			
Microphis brachyurus			X
Microphis spinachoides*		X	
Ambassidae			
Ambassis buruensis		X	
Ambassis interrupta		X	X
Parambassis altipinnis*	X		
Parambassis confinis*	X	X	X
Terapontidae			
Hephaestus transmontanus*	X	X	X
Mesopristes argenteus		X	
Kuhliidae			
Kuhlia marginata		X	
Kuhlia rupestris		X	
Apogonidae			
Glossamia beauforti	X		
Glossamia gjellerupi	X	X	X

continued

Family/Species	Mamberamo	Sepik	Ramu
Carangidae			
Caranx sexfasciatus		X	X
Lutjanidae			
Lutjanus goldiei		X	X
Sciaenidae			
Pseudosciaena soldado		X	X
Mugilidae			
Liza macrolepis		X	
Liza melinoptera		X	X
Liza tade		X	
Mugilid sp.	X		
Eleotridae			
Bunaka gyrinoides			X
Butis amboinenis		X	X
Eleotris aquadulcis*		X	X
Eleotris fusca			X
Eleotris melanosoma		X	
Hypseleotris guntheri		X	
Giurus margaritaceus	X	X	X
Mogurnda aurofodinae*	X	X	
Mogurnda nesolepis	X	X	X
Mogurnda sp.*		X	
Ophiocara porocephala		X	
Oxyeleotris fimbriata	X	X	X
Oxyeleotris heterodon*	X	X	X
Gobiidae			
Eugnathogobius tigrellus*	X		
Glossogobius bulmeri	X	X	
Glossogobius coatesi*		X	X
Glossogobius giurus		X	X
Glossogobius koragensis*	X	X	X
Glossogobius torrentis*		X	X
Glossogobius sp.*			X
Mugilogobius fusculus*		X	
Redigobius bikolanus		X	
Stenogobius laterisquamatus		X	X
Zappa confluentus			X
Gobioididae			
Brachyamblyopus urolepis		X	X
Taenioides anguillaris			X

Family/Species	Mamberamo	Sepik	Ramu
Introduced fishes			
Cyprinidae			
Cyprinus carpio	X	X	X
Barbodes gonionotus	X		
Puntius orphoides	X		
Clariidae			
Clarias batrachus	X		
Salmonidae			
Oncorhynchus mykiss			X
Poeciliidae			
Gambusia affinis		X	X
Cichlidae			
Oreochromis mossambica	X	X	X
Channidae			
Channa striata	X		
Total species	35	57	54

Appendix 26

Distribution of amphibians and reptiles in the Yongsu area (0–400 m asl), Papua, Indonesia

Stephen Richards, Djoko Iskandar, Burhan Tjaturadi, and Aditya Krishar

Trails 1 and 2 were adjacent to Yemang Camp, trail 3 was in Jari forest and trail 4 ran south from Yongsu Dosoyo (see Chapter 8 for description of sites). Trail 4 was surveyed at night only.

sp. nov. indicates an undescribed species; sp. undet. indicates existing specimens or literature insufficient for identification.

Species	Trail 1	Trail 2	Trail 3	Trail 4
Hylidae				
Litoria eucnemis			X	X
Microhylidae				
Hylophorbus sp. nov.	X	X	X	
Oreophryne sp. nov.	X	X	X	
Xenorhina oxycephala	X	X	X	
Ranidae				
Platymantis papuensis	X	X	X	X
Rana daemeli				X
Rana grisea				X
Rana papua			X	X
Lizards				
Agamidae				
Hypsilurus sp. undet. nr *dilophus*			X	
Hypsilurus modestus	X			
Gekkonidae				
Cyrtodactylus mimikanus				X
Gehyra marginata	X			
Gehyra oceanica	X			
Gekko vittatus	X			
Lepidodactylus novaeguineae	X			
Nactus pelagicus	X			
Scincidae				
Emoia caeruleocauda	X	X	X	
Emoia cyanogaster	X		X	

continued

Species	Trail 1	Trail 2	Trail 3	Trail 4
Emoia jakati	X	X	X	
Emoia verecunda	X	X	X	
Emoia sp. 1 undet.		X	X	
Emoia sp. 2 undet.	X		X	
Lamprolepis smaragdina	X		X	
Lipinia sp. undet.	X			
Sphenomorphus jobiensis		X		
Sphenomorphus simus	X	X	X	
Sphenomorphus sp. 1 undet.			X	
Varanidae				
Varanus jobiensis	X			
Snakes				
Boidae				
Candoia aspera			X	
Colubridae				
Boiga irregularis	X			
Tropidonophis sp. undet.			X	
Laticaudidae				
Laticauda sp. undet.	X (on reef)			
Turtles				
Cheloniidae				
Chelonia mydas	X (beach)			
Caretta caretta	X (beach)			

Appendix 27

Frogs and reptiles recorded from three sites in the Dabra area, Papua, Indonesia

Stephen Richards, Djoko Iskandar, and Burhan Tjaturadi

sp. nov. indicates an undescribed species; sp. undet. indicates existing specimens or literature insufficient for identification.

Species	Furu	Sites Tiri	Dabra
FROGS			
Microhylidae			
Asterophrys turpicola		X	X
Austrochaperina derongo			X
Callulops robustus	X		
Choerophryne proboscidea	X	X	X
Hylophorbus sp. nov. 1	X	X	X
Hylophorbus sp. nov. 2			X
Oreophryne sp. nov. 1 "peeper"	X	X	
Oreophryne sp. nov. 2 "rattler"	X	X	X
Sphenophryne cornuta	X	X	X
Xenorhina sp. undet.		X	
Hylidae			
Litoria infrafrenata	X		
Litoria nigropunctata	X		X
Litoria sp. nov. 1 nr *arfakiana*	X		
Litoria sp. nov. 2 nr *infrafrenata*		X	
Litoria sp. nov. 3		X	
Ranidae			
Limnonectes grunniens	X		
Platymantis papuensis	X	X	X
Rana arfaki	X	X	
Rana daemeli	X	X	
Rana grisea	X	X	X
Rana garritor	X		
Total Number of Frog Species	**15**	**13**	**10**
REPTILES			
Snakes			
Acanthophis antarcticus	X	X	
Aspidomorphus muelleri	X	X	
Boiga irregularis	X		

continued

Species	Sites		
	Furu	Tiri	Dabra
Dendrelaphis calligastra	X	X	X
Leiopython albertisi			X
Morelia viridis		X	
Stegonotus cucullatus	X		
Stegonotus diehli	X		
Stegonotus modestus	X	X	
Tropidonophis montanus		X	
Turtles			
Elseya novaeguineae	X	X	
Pelochelys cantori		X	
Lizards			
Geckos			
Cyrtodactylus mimikanus	X	X	
Hemidactylus frenatus			X
Nactus pelagicus		X	X
Skinks			
Emoia caeruleocauda	X	X	
Emoia obscura	X	X	
Emoia sp. 1 undet. nr *cyclops*	X	X	
Emoia sp. 2 undet. nr *oribata*	X	X	
Emoia sp. undet.		X	
Eugongylus rufescens	X		
Glaphyromorphus unilineatus	X		
Lamprolepis smaragdina			X
Prasinohaema? sp. undet.	X		
Sphenomorphus jobiensis			
Sphenomorphus simus	X	X	X
Sphenomorphus sp. 1 undet.	X		
Sphenomorphus sp. 2 undet.		X	
Tribolonotus novaeguineae	X		
Agamids			
Hydrosaurus amboinensis	X	X	
Hypsilurus modestus	X	X	
Hypsilurus sp. nov. nr *auritus*	X	X	
Hypsilurus sp. undet.		X	
Varanids			
Varanus jobiensis	X		
Varanus indicus		X	
Crocodiles			
Crocodylus novaeguineae		X	
Total Number of Reptile Species	23	23	6
Total Number of Species	38	36	16

Appendix 28

Annotated list of noteworthy frogs and reptiles recorded from three sites in the Dabra area, Papua, Indonesia

Stephen Richards, Djoko Iskandar, and Burhan Tjaturadi

FROGS

Hylophorbus species

Two undescribed species of *Hylophorbus* (Microhylidae) were collected in the Dabra area. Both called from beneath leaves on the forest floor. *Hylophorbus* sp. 1 (adult males SVL 26.4–31.5 mm) was common at all sites around Dabra and animals with a similar call (3–4 rather melodious notes) were also collected at Yongsu. Yongsu specimens are smaller (SVL 24.2–26.6 mm) but pending further detailed examination we tentatively consider the two populations conspecific. *Hylophorbus* sp. 2 was encountered only on the ridge immediately south of Dabra Village. Males initiated calling before dusk but calling had nearly ceased within an hour after dark in clear, dry conditions. A single specimen (SVL 26.3 mm) was obtained. Both of these species are currently being described from other localities in Papua by Dr. Rainer Günther of the Berlin Museum.

Oreophryne species

Two undescribed microhylids of the genus *Oreophryne* were recorded from the Dabra area. *Oreophryne* sp. 1 is a small (male SVL 22–24 mm) arboreal species that called at night after heavy rain at Furu and Tiri. Its advertisement call is a series of "peeps" uttered from leaves between 4 and 15 m high. *Oreophryne* sp. 2 is about the same size (male SVL 21–25 mm) but its call is a short rattle uttered from concealed perches such as curled leaves, between 1 and 3 m high. A male of this species was discovered guarding a clutch of eggs glued to the underside of a leaf about 0.5 m above the ground. These two species will be described by Stephen Richards, Djoko Iskandar, and Burhan Tjaturadi.

Litoria sp. 1

A moderately small (male SVL 31–38.5 mm) green tree frog related to *Litoria arfakiana*. Found only on a small torrential tributary of the Furu River. Most of the *Litoria arfakiana* group occur at altitudes above ~ 300 m asl (Menzies and Zweifel, 1974) and the occurrence of a species at just ~ 100 m asl near Furu Camp was surprising. The Furu taxon differs from all other members of the *L. arfakiana* group in

its distinct advertisement call which is a single musical note with 3–4 pulses.

Litoria sp. 2

A large (male SVL 56 – 68 mm) green tree frog. Morphologically similar to *Litoria infrafrenata* but differs from that species in its smaller size, absence of a distinct white stripe on the lower lip, and in having a different mating call. Encountered only in a series of shallow swamps upstream from Tiri Camp, where males called from leaves about 4–6 m high.

Litoria sp. nov. 3

A small (male SVL 22.9 mm) undescribed tree frog with striking black and yellow markings on venter, fleshy webbing between fingers, and partially transparent tympanum. A single specimen was calling from ~ 3 m high in a tree during heavy rain. Previously known only from a female specimen collected in the southern foothills of the Star Mountains, PNG. Currently being described by A. Dennis and M. Cunningham.

REPTILES

Emoia species

Skinks of the genus *Emoia* were extremely abundant in sunny patches on the forest floor at Furu and Tiri. Two of the 5 species recorded from this area (Appendix 28) do not key to any known taxon but each shows some morphological similarities to a described species. Whether these specimens represent previously unknown morphological extremes of known taxa or whether they are undescribed species will require further detailed morphological and possibly biochemical studies.

Hydrosaurus amboinensis

The large and spectacular Sail-fin Lizard (*Hydrosaurus amboinensis*) is known from scattered localities on islands between the Philippines and the western tip of Papua. It has also been reported from the Vogelkop Peninsula (Aplin 1998). One juvenile that was swimming in the Furu River was collected from disturbed forest downstream of camp. Several additional animals were observed at Tiri Camp where they perched on logs adjacent to or hanging over the river. Large animals dropped into deep pools or ran into dense riparian vegetation when disturbed. Juveniles ran across the water surface on their hind legs when disturbed. Elongated scales on the toes assist this unusual form of locomotion. Our records of this species from Furu and Tiri represent a significant easterly range extension for this poorly known animal.

Hypsilurus species

Three species of forest dragons (*Hypsilurus*) were collected in the Dabra area. One of these, *H. modestus*, is abundant in rainforest throughout lowland New Guinea and was found on low vegetation, particularly at night when they slept on palm and fern fronds ~ 1–3 m high. A single specimen of a distinctive forest dragon collected at night from forest adjacent to Tiri Camp remains unidentified pending further studies. The most conspicuous forest dragon in the Dabra area was a moderately large (SVL of our largest specimen 124 mm) riparian species that appears to represent an undescribed taxon. It occurred in micro-sympatry with *Hydrosaurus amboinensis*, often basking on logs within several meters of a Sail-fin lizard.

FRESHWATER TURTLES

Two species of freshwater turtles were found in the Dabra area, the giant soft-shelled turtle (*Pelochelys cantori*) and the smaller side-necked turtle *Elseya* sp. The taxonomy of New Guinea *Elseya* is in a state of flux and the Dabra taxon is identified only tentatively as *E. novaeguinea* following Iskandar (2000). Both of these species were common in the sluggish streams around Tiri Camp and were collected for food by locals. However the soft-shelled turtle was absent from swiftly flowing and rocky streams around Furu Camp. *P. cantori* burrow into sand on the stream bed, and the rocky stream substrates and lack of deep pools around Furu are probably unsuitable for this species. In contrast several *Elseya* were collected from streams around Furu, including two juveniles that were sheltering in root mats at the water's edge.

CROCODILES

One species, the New Guinea Freshwater Crocodile (*Crocodylus novaeguineae*) was recorded at Tiri. This widespread species is found in lowland swamps and rivers across mainland New Guinea. Populations in the Mamberamo Basin have been hunted extensively and local informants suggested that over-harvesting has reduced numbers to a critically low level. One animal travelled upstream along the Tiri River, through the centre of our camp at night leaving a conspicuous trail. Local hunters tracked the animal and captured it in a deep pool upstream of camp.

LITERATURE CITED

Aplin, K. 1998. Vertebrate Zoogeography of the Bird's Head of Irian Jaya, Indonesia. *In:* Miedema, J., C. Odé and R.A.C. Dam (eds) Perspectives on the Bird's Head of Irian Jaya, Indonesia: Proceedings of the Conference Leiden, 13–17 October 1997. Amsterdam: Rodopi. Pp 803–890.

Iskandar, D.T. 2000. Turtles and crocodiles of insular Southeast Asia and New Guinea. Bandung: Institute of Technology.

Menzies, J.I. and Zweifel, R.G. 1974. Systematics of *Litoria arfakiana* of New Guinea and sibling species (Salienta: Hylidae). American Museum Novitates 2558: 1–16.

Appendix 29

Birds recorded at Yongsu, Papua, Indonesia

Pujo Setio, Paul Johan Kawatu, Irba U. Nugroho, David Kalo, Daud Womsiwor, and Bruce M. Beehler

Species	# of Days Recorded	Mean # Recorded Daily	Highest Daily Count	Comments
Casuarius unappendiculatus	4	0.4	1	includes fresh dung pile
Egretta sacra	2	0.2	1	dark morph
Haliastur indus	6	0.8	2	once on coastal rocks
Haliaeetus leucogaster	2	0.2	1	flying along coastline
Accipiter novaehollandiae	2	0.2	1	over forest
Accipiter poliocephalus	2	0.2	1	one netted
Pandion haliaetus	1	0.1	1	along coast
Talegalla jobiensis	8	2	4?	nesting completed
Sterna bergii	1	0.2	2	at sea by bay
Actitis hypoleuca	2	0.2	1	river and beach
Ptilinopus superbus	8	1.1	2	calling in forest canopy
Ptilinopus ornatus	4	1	4	calling in forest canopy
Ptilinopus magnificus	7	1	2	heard, not seen
Ducula zoeae	7	0.9	3	heard often, seen once
Ducula bicolor	6	11.6	28	by coast and over bay
Ducula pinon	1	0.1	1	seen once over bay
Macropygia amboinensis	4	0.4	1	heard rarely
Reinwardtoena reinwardtii	9	1.3	3	seen daily, vocal
Chalcophaps stephani	1	0.1	1	one netted
Trichoglossus haematodus	10	18.1	31	abundant in flocks
Lorius lory	10	4.5	12	usually in pairs
Charmosyna rubronotata	2	0.7	5	scarce, in forest canopy
Micropsitta pusio	9	3.5	8	common and vocal, in forest
Probosciger aterrimus	7	0.9	2	vocal
Cacatua galerita	7	1.9	5	vocal
Eclectus roratus	3	0.4	2	uncommon
Geoffroyus geoffroyi	2	1.7	15	seen only twice
Cacomantis variolosus	4	0.4	1	
Cacomantis castaneiventris	1	0.1	1	heard once
Microdynamis parva	3	0.3	1	heard in forest

continued

Species	# of Days Recorded	Mean # Recorded Daily	Highest Daily Count	Comments
Scythrops novaehollandiae	6	0.7	2	vocal
Centropus menbeki	5	0.6	2	vocalization only
Collocalia esculenta	9	3.9	10	open areas and coast
Collocalia cf. *vanikorensis*	7	6.5	20	over bay
Ceyx lepidus	7	0.9	2	heard in forest
Dacelo gaudichaud	5	0.5	1	heard in forest
Melidora macrorrhina	4	0.7	2	heard at dusk, netted twice
Halcyon saurophaga	3	0.8	5	rocky coast and beach
Halcyon sancta	4	0.4	1	uncommon, only around camp
Halcyon tototoro	8	1	3	forest interior
Merops ornatus	5	0.7	2	open areas, beach
Aceros plicatus	10	5.4	8	vocal and conspicuous
Pitta erythrogaster	1	0.2	2	heard once
Hirundo tahitica	4	0.7	3	
Coracina boyeri	8	1.3	3	identified by voice
Coracina melaena	5	0.6	2	in mixed flocks
Lalage atrovirens	5	1.1	4	learned song by day #5
Ptilorrhoa caerulescens	8	0.9	2	forest interior
Crateroscelis murina	8	0.9	2	forest interior
Sericornis spilodera	6	1.5	4	in flocks, vocal
Gerygone chrysogaster	8	1	2	common
Gerygone chloronotus	9	1.1	2	common
Gerygone palpebrosa	6	0.6	1	uncommon
Rhipidura leucothorax	1	0.1	1	heard behind village
Rhipidura rufidorsa	1	0.2	2	two netted
Rhipidura rufiventris	6	0.9	3	
Dicrurus bracteatus	10	1.2	3	
Monarcha chrysomela	2	0.2	1	
Arses telescophthalmus	7	1.1	3	
Machaerirhynchus flaviventer	1	0.1	1	
Monachella muelleriana	1	0.3	3	waterfall and Kali Dan Yamo
Microeca flavovirescens	2	0.2	1	requires confirmation
Pachycephala simplex	1	0.1	1	heard in forest once
Pitohui kirhocephalus	10	4.5	6	common
Pitohui ferrugineus	5	1.3	5	
Philemon meyeri	7	1.3	4	
Philemon buceroides	10	3.4	6	very common and vocal
Melilestes megarhynchus	6	0.7	2	elusive, in forest
Xanthotis flaviventer	8	0.9	2	uncommon
Meliphaga aruensis	4	0.5	2	
Meliphaga analoga	10	2	5	tentative identification
Nectarinia aspasia	10	2.3	5	common, especially at forest edge

continued

Species	# of Days Recorded	Mean # Recorded Daily	Highest Daily Count	Comments
Nectarinia jugularis	2	0.2	1	uncommon, openings
Dicaeum pectorale	4	0.9	3	
Toxorhamphus novaeguineae	9	1.9	4	forest interior
Oedistoma iliolophum	6	0.9	2	fairly common
Oedistoma pygmaeum	9	2.5	5	common and vocal
Melanocharis nigra	8	1.9	5	
Zosterops atrifrons	6	0.8	2	vocal, not in flocks
Aplonis metallica	9	14.8	50	active, colony behind village
Mino dumonti?	1?	0.1	1	vocalization only
Cracticus quoyi	10	2.4	5	vocal and common
Cracticus cassicus	5	0.7	3	requires confirmation
Manucodia sp.	4	0.5	1	elusive, vocalization only
Ptiloris magnificus	10	1.7	2	vocal
Cicinnurus regius	8	1	2	
Cicinnurus magnificus	1	0.1	1	heard once in forest across bay from camp
Paradisaea minor	10	4.5	10	common and vocal
Gymnocorvus tristis	9	3.3	6	
Corvus orru	9	1	2	vocal

Appendix 30

Forest trees at two bird census points, Yongsu, Papua, Indonesia

Pujo Setio, Paul Johan Kawatu, Irba U. Nugroho, David Kalo, Daud Womsiwor, and Bruce M. Beehler

Trees at Census Point 3 (70 m asl) Species	Family	Present at Census Point 4 (40 m asl)
Actinodacna sp.	Lauraceae	x
Aglaia sp.	Meliaceae	x
Buchanania sp.	Anacadiaceae	x
Canarium oleosum	Burseraceae	x
Celtis sp.	Ulmaceae	x
Cerbera floribunda	Apocynaceae	x
Cynometra rhamnifolia	Fabaceae	x
Calophyllum inophyllum	Clusiaceae	x
Decaspermum fruticosum	Myrtaceae	x
Dillenia sp.	Dilleniaceae	x
Diospyros sp.	Ebenaceae	x
Elaeocarpus sp.	Elaeocarpaceae	x
Eugenia sp.	Myrtaceae	x
Fagraea sp.	Loganiaceae	x
Glochidion sp.	Euphorbiaceae	x
Gonocarium	Icacinaceae	-
Intsia bijuga	Leguminosae	-
Dracontomelon sp.	Anacardiaceae	x
Horsfieldia silvestris	Myristicaceae	x
Litsa sp.	Lauraceae	x
Manilkara fasciculata	Myrtaceae	x
Macaranga aleuritoides	Euphorbiaceae	x
Palaquium amboinensis	Sapotaceae	x
Pometia sp.	Sapindaceae	x
Syzygium sp.	Myrtaceae	x
Teysmaniodendron sp.	Urbinaceae	-
Podocarpus blumei	Podocarpaceae	x
Gnetum gnemon	Gnetaceae	x
Knema tomentela	Myristicaceae	x
Dysoxylum sp.	Meliaceae	x
Campnosperma gravipetiolata	Anacardiaceae	-
Pterocarpus indica	Papilionaceae	-

Appendix 31

Birds recorded from the Mamberamo/ Idenburg river basins, Papua, Indonesia

Bas van Balen, Suer Suryadi, and David Kalo

Abundance Categories
A- abundant
C- common
F- fairly common
U- uncommon
R- rare
X- present
* - endemic to New Guinea

Evidence
S- sighting
V- voice record
R- sound (MD) recording
T- tracks, feathers, live specimens
other than mist netted
M- mistnetted and photographed
P- preserved skins

Local names
D- Dasigo
A- Airo
K- Kaowerawetj

Site	FURU	DABRA	BUARE	TIRI		PRAUWEN	PIONIER	BERNHARD	
Survey	RAP	RAP	RAP	RAP	Evidence for RAP records	van Heurn 1920–21	van Heurn 1920–21	van Eechoud, 1939	Archbold 1938–39
Species									**Local Names**
Casuarius unappendiculatus	C			C	VT	X			X	ku (D), kwa (A)
Phalacrocorax sulcirostris			A		S				X	krerij (D), kerio (A)
Phalacrocorax melanoleucos			C		S				X	krerij (D), kerio (A)
Anhinga novaehollandiae			U		S				X	krerij, butukare (D), keriyo (A)
Threskiornis molucca								X	R	wa (D), pawai (K)
Ardea sumatrana	R		C		S		X		U	terijkuj (D), gugu (A)
Casmerodius albus			C		S	X			F	terijkuj, kogare (D)
Egretta picata			C		S			X	U	kudaso (D), kujaso (A), pitemu (K)
Egretta intermedia			U		S				C	frijei, kudaso (D), kogare (D, A)
Egretta garzetta			C		S				C	frijei, kudaso (D), kogare (D, A)
Butorides striatus			U		S				F	terijkuj, ko (D)
Nycticorax caledonicus							X		R	kaso (D, A)
Zonerodius heliosylus						X	X		U	frege (D, A)
Ixobrychus sinensis									C	

continued

Site	FURU	DABRA	BUARE	TIRI	Evidence for RAP records	PRAUWEN	PIONIER		BERNHARD	
						van Heurn 1920–21	van Heurn 1920–21	van Eechoud, 1939	Archbold 1938–39	
Survey	RAP	RAP	RAP	RAP						
Species										**Local Names**
Dupetor flavicollis							X		F	kogi (D,A)
Pandion haliaetus			U		S				R	tufwa (D)
Aviceda subcristata		U	U		S	X	X	X	U	tuso, tufidzo (D), duaso (A)
Macheiramphus alcinus				R	SR					
*Henicopernis longicauda	U				S		X	X	U	tuso (D), duaso (A)
Haliastur sphenurus			U		S				F	fuso (D), eyjakoni, dukaruy (A)
Haliastur indus	C	C	C	C	S	X			C	kwaragotia (D), burerey, suruy (A)
Haliaeetus leucogaster		U			SR				F	ruj (D), eyjasudu, duaso (A)
Accipiter novaehollandiae	?R				S				F	tufra, tufoga (D), duaso (A)
*Accipiter poliocephalus				U	S			X		tufwa (D), sosonokan (K)
Accipiter cirrhocephalus		U			S					
*Accipiter doriae							X			
Hieraaetus morphnoides	U				S				R	tuso (D), duaso (A)
Falco berigora							X			
Dendrocygna guttata						X			C	turebi (D), riydu (A)
Tadorna radjah			U		S					krarigo (D)
Megapodius freycinet	C			U	R		X	X	X	fore, kwiyeke (D), iyo (A), payi (K)
*Talegalla jobiensis	C	C		C	R	X		X	X	fore (D), kiserdi, kisgai (A), koreta (K)
Rallina tricolor							X		X	
*Megacrex inepta									X	ijdege (D), eijage (A)
Eulabeornis plumbeiventris						X				fikagu (D), kigo (A)
Amaurornis olivacea		C			V		X		X	kri (D), kigo (A)
Porphyrio porphyrio									X	
Poliolimnas cinerea									F	
Actitis hypoleucos			U		S	X	X	X	X	wijtu (D), wijdu (A), sani (K)
Irediparra gallinacea									X	
Himantopus leucocephalus			U		S				R	
Sterna albifrons							X		F	tutbe (D), duture (A)
Ptilinopus magnificus	C	C		C	SR		X	X	X	towa, ijderi (D), noma (K)
*Ptilinopus ornatus				R	S					ijderi (D), egeri (A)
*Ptilinopus perlatus							X	X		cit (K)
*Ptilinopus aurantiifrons									X	
Ptilinopus superbus	C			A	SR			X	X	ijderi (D), egeri (A), ehey (K)
*Ptilinopus pulchellus								X		kogobor, kokobura (K)
Ptilinopus coronulatus	C			A	SR	X	X		X	ijderi (D), egeri (A)
*Ptilinopus iozonus	A			A	SR		X	X	X	ijderi (D), egeri (A)

continued

Site / Species	FURU	DABRA	BUARE	TIRI	Evidence for RAP records	PRAUWEN	PIONIER		BERNHARD	Local Names
Survey	RAP	RAP	RAP	RAP		van Heurn 1920–21	van Heurn 1920–21	van Eechoud, 1939	Archbold 1938–39	
Ptilinopus nanus									X	
Ducula rufigaster	U			C	SR	X		X	X	kerijweij (D), sowetsa, geriweig (A), idis (K)
Ducula pinon	A			A	SR	X	X	X	X	kugdo (D), aurjo (A), amesin (K)
Ducula muellerii	R				S		X	X	R	kerijweij (D), amesin (K)
Ducula zoeae	C			A	SR	X·	X	X	F	imai (D), imayhi (A), titeris, dideris (K)
Gymnophaps albertisii				C	S?R	X			X	krikudo (D), fratu (A)
Macropygia amboinensis	U			U	SR	X		X	F	kufo (D, A), waremes (K)
Macropygia nigrirostris									F	
Reinwardtoena reinwardtii	C			C	SR			X		kufo (D, A), sitikwa (K)
Henicophaps albifrons						X	X	X	X	piripide, tiretis, dudutia (K)
Chalcophaps stephani							X		X	ijeri (D)
Trugon terrestris						X	X			ocej (K)
Gallicolumba rufigula				U	SV	X			X	brude (D)
Otidiphaps nobilis	U				V	X	X		X	sweigde, sgweide (D)
Goura victoria	C			C	S	X	X	X	X	tobig (D), dukure (A), maris (K)
Chalcopsitta duivenbodei	A	U		A	SR	X	X	X	X	awiki, suri (D), di (A), kwesij (K)
Trichoglossus haematodus	C				S	X	X	X	X	suri, bogaro (D), di (A), kwisir (K)
Pseudeos fuscata	C			C	S				X	suri (D), di (A)
Lorius lory	C	C		C	SR	X	X	X	F	wijdogwei (D), wijeysuru (A), uwera (K)
Charmosyna placentis			?R	?R	S					kujsuri (D)
Charmosyna pulchella									X	
Charmosyna josefinae									X	
Psittaculirostris salvadorii	R				SR				R	kerejtri (D), hayi (A)
Micropsitta pusio						X	X	X	X	keretri (D), cisidis (A)
Opopsitta diophthalma	R			?R	S					
Probosciger aterrimus				R	S	X	X	X	X	tijwea (D), dijwia (A), awehi (K)
Cacatua galerita	A	A	C	A	SR	X	X	X	C	wia (D), ase (A), pawuy (K)
Psittrichas fulgidus				U	S			X	X	ayi, aij (D), ai, hayi (A), ciahari (K)
Eclectus roratus	A			C	S	X	X	X	X	frijei, fri(k)dog (D); feriyji (A), sopis (K)
Geoffroyus geoffroyi	C	C		C	SV	X	X	X	C	braro (D), bagaro (A) male: bakan, pagan; female: tagen (K)
Alisterus amboinensis								X		
Alisterus chloropterus								X		musiyer (K)

continued

Site	FURU	DABRA	BUARE	TIRI	Evidence for RAP records	PRAUWEN van Heurn 1920–21	PIONIER van Heurn 1920–21	van Eechoud, 1939	BERNHARD Archbold 1938–39	
Survey	RAP	RAP	RAP	RAP						
Species										**Local Names**
*Loriculus aurantiifrons		?R			S					keretri (D)
Cacomantis variolosus	C	C		C	S		X		X	suwi (D, A)
Rhamphomantis megarhynchus						X				
Chrysococcyx russatus									X	
*Caliechthrus leucolophus									X	
*Microdynamis parva	C			C	R					
Eudynamys cyanocephala	C			C	R		X			male: kweiro, female: kwaguriko (D)
Scythrops novaehollandiae	R				S			X		
*Centropus bernsteini						X			F	
*Centropus menbeki	C			C	SR	X	X	X	F	frigik (D), orufji (A), sim (K)
*Ninox theomacha				U	R					tobig, tu'ku (D)
Podargus papuensis				C	V	X	X	X	X	boujsa (D, A), udud (K)
Podargus ocellatus	U			U	M				X	boujsa (D, A)
*Aegotheles bennettii									X	
Eurostopodus mystacalis									X	
*Eurostopodus papuensis									X	
Caprimulgus macrurus		C			R					twu, tgu (D)
*Aerodramus papuensis									X	frobare (D, A)
Aerodramus vanikorensis	C	C		C	S					frobare (D, A)
*Mearnsia novaeguineae		A		R	S				C	frobare (D, A)
Hirundapus caudacutus	R				S					frobare (D, A)
Hemiprocne mystacea		C			S		X	X	X	frobare (D, A), atenaten (K)
Alcedo azurea	U			C	STM	X	X		X	firi (D), setyaka (A)
Alcedo pusilla	R				M				F	firi (D), seiyaka (A)
Ceyx lepidus	C			C	M	X	X	X	X	firi (D), seiyaka (A), tes (K)
*Dacelo gaudichaud	A	C		C	SR	X	X	X	X	wrigti, wrijdik (D), kwesudo (A), karosen (K)
*Clytoceyx rex		U		R	R	X		X	X	kotigde (D), kowis (K)
*Melidora macrorrhina	?U			C	R	X	X	X	F	sotu (D), sodu (A)
Halcyon torotoro	C				R		X	X		sotu, sigei (D), sotu (A), abidareij (K)
Halcyon sancta							X	X	X	feiydawrijdi (D), dirakwesudo (A)
Tanysiptera galatea				C	M	X	X	X	X	sigeg (D), sigeyk (A), sekarara (K)
Merops ornatus			U		S		X		C	trarikwa, trorupa (D), kwero (A)
Eurystomus orientalis	C	C			S	X	X		X	srigt(e)go (D), sarijare (A)
Rhyticeros plicatus	A			A	S			X	X	ebeij (D), keijra (A), muk (K)

continued

Site	FURU	DABRA	BUARE	TIRI	Evidence for RAP records	PRAUWEN	PIONIER	BERNHARD	
Survey	RAP	RAP	RAP	RAP		van Heurn 1920–21	van Heurn 1920–21	van Eechoud, 1939	Archbold 1938–39
Species									**Local Names**
Pitta erythrogaster	U			U	V		X	X	X	krwewejkori (D), okweo (D, A), kogora (K)
Pitta sordida									X	krwewejkori (D)
Hirundo rustica			R		S					frobare (D)
Hirundo tahitica			U		S		X		X	frobare (D)
Cecropis nigricans		C			S		X			
Motacilla cinerea					/	X				
Coracina papuensis	C	C		U	S		X		F	fiso, krubia, bidereid (D)
Coracina boyeri									X	
Coracina incerta									X	
Coracina schisticeps	U				SR		X		X	krubia (D)
Coracina melas	C			C	SR	X	X		X	krubia, kabi (D)
Campochaera sloetii	C			U	SR				X	kwawuya, kwerewe, traritu (D)
Lalage atrovirens	U			U	SR		X		X	suwi (D)
Ptilorrhoa caerulescens	R				M			X	X	rideraitaritobij (D), marisaya, marisiyagai (K)
Pomatostomus isidorei				U	S		X		X	ewai (D)
Malurus alboscapulatus			R		S		X	X	C	ijijemayis (K)
Malurus cyanocephalus	R				S	X				kbitu (D)
Malurus grayi							X			kbitu (D)
Acrocephalus orientalis									X	
Crateroscelis murina							X	X	X	
Sericornis spilodera									C	
Gerygone chloronotus				U	R				R	suwi (D)), swi (D, A)
Gerygone palpebrosa							X	X		swi (A), beberisopis (K)
Gerygone chrysogaster	C			C	SR	X	X		X	suwi (D), swi (D, A)
Gerygone magnirostris			C	C	SR	X	X		C	suwi (D), swi (D, A)
Monarcha guttula	U	C		C	S	X	X	X	X	cakuretitecem (immature), p(r)obolisis (K)
Monarcha rubiensis				?R	S		X			
Monarcha chrysomela	U				SR	X	X	X	X	editetan (K)
Monarcha manadensis						X		X	X	pobolisis (K)
Arses insularis									X	
Arses telescophthalmus	C			U	SR	X	X	X	X	etawegi (K)
Piezorhynchus alecto	U				SR		X	X	X	foreto (D), foredu (A), parowi (K)
Peltops blainvillii	C			C	SR	X	X	X	X	kwero (D), katun (K)
Rhipidura leucophrys			C		S	X	X	X	F	troujda (D), sijsokwriri (A), boborokes (K)
Rhipidura rufiventris	C	C		C	SR	X	X	X	X	sooli (D), sijsokwriri (D, A), orotow (K)

continued

Site	FURU	DABRA	BUARE	TIRI	Evidence for RAP records	PRAUWEN	PIONIER		BERNHARD	
Survey	RAP	RAP	RAP	RAP		van Heurn 1920–21	van Heurn 1920–21	van Eechoud, 1939	Archbold 1938–39	
Species										Local Names
*Rhipidura threnothorax	U			C	SR				X	sijsokwriri (D, A)
*Rhipidura hyperythra								X		sijsokwriri (D, A), enenokwehere, momosihubyao (K)
*Rhipidura leucothorax	C				SR	X	X		X	sijsokwriri (D, A)
*Rhipidura rufidorsa	U				S	X				sijsokwriri (D, A)
*Microeca flavovirescens	R				S			X		kagarabyej (K)
Monachella muelleriana								X		ononahet, orotow (K)
*Poecilodryas hypoleuca						X	X		F	orotow (K)
*Poecilodryas brachyura								X		orotow (K)
*Pachycephala hyperythra									X	
Pachycephala simplex				?U	R		X	X	X	editetan (K)
Colluricincla megarhyncha		C		C	SR	X				
*Pitohui kirhocephalus				?R	S	X	X		X	tutog, tutoij (D), dutoij (A)
*Pitohui cristatus									X	
*Pitohui ferrugineus	A			C	SRM	X	X	X	X	tutog, tutoij (D), aciu (K)
*Melanocharis nigra	C			U	SM	X	X	X	X	suwi (D), ijijemayis (K)
*Dicaeum pectorale	U	C		U	S			X		suwi (D)
Nectarinia jugularis									X	
Nectarinia aspasia	C	C		C	S	X			X	suwi, tubiki (D)
*Melilestes megarhynchus	U	C		C	SR	X	X	X	X	tow, bago (D), fisari (A), kwasir (K)
*Toxorhamphus novaeguineae	R			R	S		X	X	X	kwekwesa (K)
*Oedistoma iliolophum	C				SR					
*Oedistoma pygmaeum				R	S					
*Glychichaera fallax									X	
Myzomela eques						X			X	
*Meliphaga montana				R	S					wasire (D)
*Meliphaga flavirictus									X	
Meliphaga aruensis	C				M	X			X	wasire (D)
*Meliphaga analoga						X	X		X	wasire (D), sorei (K)
Xanthotis chrysotis	C			C	S	X	X	X	X	arokwes (K)
*Pycnopygius stictocephalus							X		X	
*Pycnopygius ixoides		?U			S	X	X		X	pobokwaj (K)
*Philemon brassi									X	
*Philemon meyeri		?U		?U	S	X			X	(m)bogo (D), obogo (A)

¹ collected at Batavia Bivak

continued

Site	FURU	DABRA	BUARE	TIRI	Evidence for RAP records	PRAUWEN	PIONIER		BERNHARD	
						van Heurn 1920–21	van Heurn 1920–21	van Eechoud, 1939	Archbold 1938–39	
Survey	RAP	RAP	RAP	RAP						
Species										Local Names
Philemon novaeguineae	A	C	C	C	SR	X	X		X	(m)bogo (D)
Lichmera alboauricularis			U		ST				C	
Lonchura tristissima							X		X	
Lonchura grandis		A			S		X[1]		C	ijfariki, bitu (D), sasu (A)
Aplonis cantoroides		R			S			X	U	srijsorij (D), arukwes, matagori (K)
Aplonis metallica	R		R		S	X	X	X	C	srijsorij (D), arena (K)
Mino anais	C			C	SR			X	X	kokba, saukride (D), kokwa (A), tegej (K)
Mino dumontii	C	U		C	SR	X	X	X	F	kokba (D), kokwa (A)
Oriolus szalayi	U		U	U	SR		X		F	sri (D)
Dicrurus bracteatus	A			C	SR	X	X	X	X	ketire D), sisireki (A), teter (K)
Artamus leucorynchus	U	C	U	U	S	X	X	X	F	tibia (D), teraij (A), segeret, sekerej (K)
Cracticus cassicus				C	SV	X	X	X	C	kugtbe, kujtbe (D), kobu (A), polis (K)
Cracticus quoyi	U			C	R	X			X	kwagumko (D), kobujkeke (A)
Ailuroedus buccoides				C	SR		X	X	X	igerag, bijdraij (D)
Manucodia atra	R			?R	S	X	X	X	X	eijdafogi (D)
Manucodia jobiensis	U				S	X		X	X	eijdafogi (D), sotiwer (K)
Manucodia chalybata									X	
Ptiloris magnificus	?R	C		U	V			X	X	wa, (t)owa (D), towa (A), abosis (K)
Seleucidis melanoleuca				R	SR		X	X	X	twougtigare, tgoutiyari (D), dewisuydu (A), abosis (K)
Epimachus bruijnii							X	X		oraris (K)
Cicinnurus regius	C	C		C	SR	X	X	X	X	ka (D, A), anenisis (K)
Paradisaea minor	C	C		U	SR	X	X	X	X	tiyare (D), foruseri (A), kaji (K)
Corvus tristis	U	U			SR			X	X	ke (D, A)
Totals	**87**	**39**	**29**	**87**	**143**	**67**	**100**	**87**	**164**	

Appendix 32

Annotated list of noteworthy birds known from or expected to occur in the Dabra area

Bas van Balen, Suer Suryadi, and David Kalo

The global status (vulnerable, data-deficient, near-threatened) of the following species is based upon Stattersfield and Capper (2000); restricted range species (RRS) are those associated with the North Papuan lowlands Endemic Bird Area (EBA; Stattersfield et al., 1998).

Northern Cassowary (*Casuarius unappendiculatus*)

Local informants reported that hunting pressure on cassowaries is high in the immediate vicinity of Dabra village and an adult was killed near our Furu camp during the survey. Despite this, cassowaries appeared to be common in the forests around Dabra and footprints were regularly encountered on trails and creek banks near our Furu and Tiri campsites. They are reportedly kept as pets in Dabra.

Bat Hawk (*Macheiramphus alcinus*)

A single immature bird was seen and tape-recorded near the Tiri campsite on 11 September. This is only the second, and best-documented, record of the species in Papua. The only previous sighting was by Glynn (1995; without published details) at Kobakma, less than 100km SSE of our survey plot.

Victoria Crowned Pigeon (*Goura victoria*) RRS

Crowned Pigeons were common in the immediate vicinities of the Furu and Tiri campsites although their density decreased within a few days of camp occupancy, possibly due to the noise and disturbance.

Palm Cockatoo (*Probosciger aterrimus*)
Near-threatened

One or two birds were seen near our Tiri camp site.

Salvadori's Fig-Parrot (*Psittaculirostris salvadorii*)
Vulnerable, RRS

The pet trade is believed to have had a major impact on populations of this species. However there was no evidence of trade in this species at Dabra and little evidence of bird trapping in the area. The only caged birds seen were several Sulphur-Crested Cockatoos and Rainbow Lorikeets that were held captive or were being shipped out of Dabra.

Brown Lory (*Chalcopsitta duivenbodei*) RRS

This restricted-range species was common around our sites. It was usually seen flying over the forest in noisy flocks. Its rather dull plumage makes this species less popular in the bird trade. We obtained voice recordings of this species.

Vulturine Parrot (*Psittrichas fulgidus*)
Vulnerable

Three birds were seen near a fruiting tree close to the Tiri camp site. Birds were seen flying over camp the previous day.

The following noteworthy species have been collected in the Mamberamo area by previous expeditions but were not recorded during the present survey. More intensive surveys may reveal their presence.

Doria's Hawk (*Accipiter [Megatriorchis] doriae*)
Near-threatened

A skin was collected at Pionier Bivouac by van Heurn, in June 1920.

Brass's Friarbird (*Philemon brassi*)
Near-threatened, RRS

Collected by the Archbold expedition (1938–1939). Although we visited a lagoon that—according to published descriptions of its habitat—was expected to contain the friarbird, we did not find a single bird. More intensive searching of the many lagoons dispersed along the Mamberamo and Idenburg rivers may reveal additional populations of this species.

Pale-Billed Sicklebill (*Epimachus bruijnii*)
Near-threatened, RSS

Two skins were collected by van Heurn at Pionier Bivouac during June–December 1920 and another two skins were collected at the same locality by van Eechoud during July-November 1939. This bird of paradise is a poorly known species that typically inhabits the canopy of floodplain forests.

LITERATURE CITED

Glynn, W.F. 1995. Bat Hawk (*Macheiramphus alcinus*) sighting from Kobakma, Irian Jaya, Indonesia on 8th December 1990. Muruk. 7: 112–123.

Stattersfield, A.J., M.J. Crosby, A.J. Long, and D.C. Wege. 1998. Endemic Bird Areas of the World. Priorities for Biodiversity Conservation. Birdlife Conservation Series No. 7.

Stattersfield, A.J. and D.R. Capper. 2000. Threatened birds of the world. Cambridge: BirdLife International.